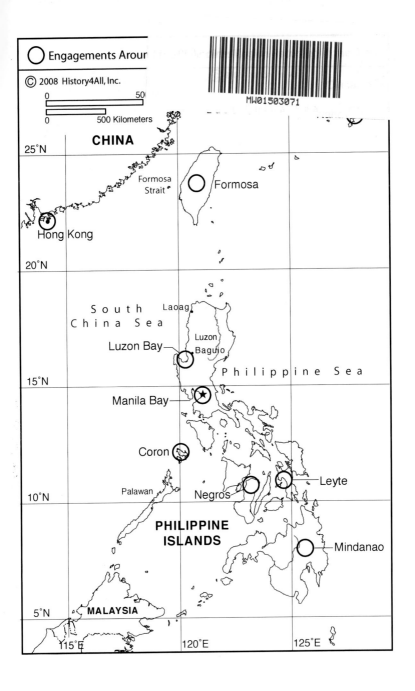

Engagements Arour

© 2008 History4All, Inc.

0 — 50

0 — 500 Kilometers

CHINA

25°N

Formosa Strait

Formosa

Hong Kong

20°N

South China Sea

Laoag

Luzon Bay

Luzon

Baguio

Philippine Sea

15°N

Manila Bay

Coron

Palawan

Negros

Leyte

10°N

PHILIPPINE ISLANDS

Mindanao

5°N   MALAYSIA

115°E   120°E   125°E

MW01503071

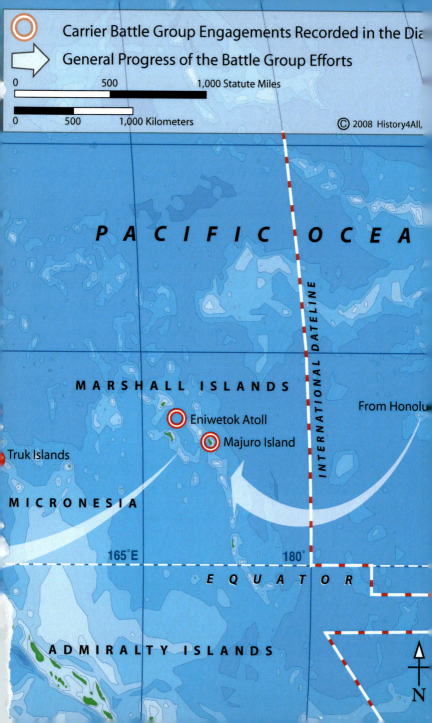

Naomi Sokol Zeavin and Carmen's sister Barbara Franceschelli.
Of her late brother Carmen, Barbara says:

> *I am happy that Carmen's diary is being published so people can
> better understand the emotional costs of a war. With this diary,
> perhaps we can do a better job of understanding the effect of
> the experience. And since Carmen died young, this book at least
> enlarges the life he had and the lives of others who served with
> him.*

Carmen's Secret Diary

Carmen's Secret Diary

Aboard the USS *Hornet* (CV-12) in 1944

Naomi Sokol Zeavin

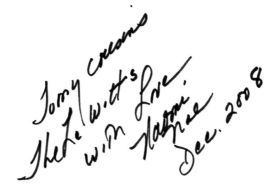

History4All, Inc.
Fairfax, Virginia

Carmen's Secret Diary:
Aboard the USS *Hornet* (CV-12) in 1944
Naomi Sokol Zeavin

© 2008 by Naomi D. Zeavin

Published by History4All, Inc.
Post Office Box 1126
Fairfax, Virginia 22038
history4all-books@live.net

ISBN 978-1-934285-02-2
Library of Congress Control Number 2008920915

Printed in the United States of America
First Edition
First Printing, September 2008
Second Printing, October 2008

Maps and Cover Design by Lyonshare Studios LLC, Mason Neck, Virginia.
Compositing and Manufacturing by Baldersons Inc., Chantilly, Virginia.

Dedicated to my children

*Gerald, Bradford, Jill, Spencer,
and to my five grandchildren
and two step-grandchildren.*

Table of Contents

1945 Left Ulithi 0745 had G.Q. while leaving the island, three full deck loads took off altogether we had four launches the last one was the 4th patrol. Reed and I fixed six turis the air group did very well I watched most of them take off the Hornet for the first time. For practice instead of water bombs they used 100 pound bombs rockets, and gunnery practice on a shee sleel about three hundred yards away from the ship it was really something to watch. Had ship gunnery practice also night practice 5" 40 & 20 M.M. I think the air group is going to be tops.

A page in Carmen Franceschelli's World War II diary.

Preface

While in my hometown of New Britain, Connecticut, doing research for a forthcoming book about the huge wartime contributions made by the town's many manufacturing plants and workers during World War II, I met with Barbara Franceschelli. Her mother had worked at my family's coat manufacturing plant. Barbara told me about her brother Carmen's Navy service and showed me his diary. I read it, and then asked permission to transcribe and publish it, which she kindly granted.

Every day for 366 days Carmen Franceschelli secretly wrote in this small book while aboard the aircraft carrier USS *Hornet* (CV-12). A sample page in his actual handwriting is reproduced on the page to your left. I think it is good handwriting for having been written at night under the covers with only a flashlight to see by.

For two reasons, no photographs or other illustrations have been inserted within the diary text. First, there already are excellent books published which contain many photographs showing the USS *Hornet* during World War II. *The United States Ship Hornet: First War Cruise, 1943-1945* is filled with official photographs of shipboard life that were taken aboard the *Hornet* during the war. Another is *Grey Ghost* by Lee W. Meridith. This book, published in 2001, covers the entire history from the earliest *Hornet* (1775) up through providing a self-guided tour of the USS *Hornet*, now anchored in Alameda, California, as a floating museum (www.uss-hornet.org).

The second reason is that Carmen's words paint clear and colorful pictures of the actions aboard ship, and in the water and sky surrounding him. In transcribing the diary, the goal was to not change his words and spellings very much. However, as Carmen tended to write run-on sentences, punctuation (mostly periods, sometimes commas or colons) have been added to make it easier for you to read. The result often times is sentence fragments. The meanings come through more clearly, though.

His use of capital letters is somewhat random. One word he <u>always</u> capitalizes, however, is "Captain." Carmen obviously respected the boss of his ship. In most words with double consonants—like "spotting" of planes on the flight deck—he uses only one consonant ("spoting"), but you will get used to that as you read along. The same with his using an ampersand (&) instead of "and" (which he frequently spelled as "an" when he did write it out).

Military time (24-hour clock) has been used consistently throughout the transcription, although Carmen switched between military time and civilian time. At the beginning of the diary, he added the day of the week to the date; I've continued that throughout the entire book. Once he learned about crossing the International Date Line, he started referring to the "old time" and the "new time" so two days are noted. (See Appendix A for photographs of the Shellback Initiation.)

To help you follow the ship's travels, two maps are included within this book. One is of the Philippines Region. The other is a larger, fold-out map showing the Pacific Theatre.

It is unfortunate that we will never know what Carmen would

think of his diary being published for everyone to read his thoughts about being out in the Pacific Ocean for an entire year starting when he was only 19 years old. He survived the war and returned home to Connecticut. In 1956 he moved south for his health, transferring his job with the U.S. Postal Service from New Britian to Florida. There he died in an automobile accident at age 33. We are grateful to his family for being willing to share Carmen's story—the common sailor's story—with us after all these years.

Falls Church, Virginia
August 2008

Naomi Sokol Zeavin

Secretary of the Navy Frank Knox watches as Annie Reid Knox christens the USS *Hornet* (CV-12) at Newport News, Virginia, on August 30, 1943.

Introduction

Born on June 3, 1924, like many young men of his generation Carmen Franceschelli had never traveled far from his home in Connecticut. He lived with his family. His father, Vincent Franceschelli, owned Jimmie's Barber Shop. His mother, Rachel D'Agostino Franceschelli, sewed for Sokol Brothers Coat Factory. Both parents were Italian immigrants. Carmen had one brother and one sister. While reading the diary, you will come across references to his brother Anthony and "sis" Barbara.

He graduated from New Britain High School on June 25, 1943. His life centered on work and family. Mechanical drawing was Carmen's career choice, and a fine mind with attention to detail surfaces again and again in the pages of his diary. Working as a draftsman for six months at Corbin Screw, he then enlisted in the U.S. Navy. *The Commissioning Muster Roll of the Crew of the U.S.S. Hornet CV-12 for the Commissioning Date 29 November 1943* shows him aboard the *Hornet.*

The USS *Hornet* (CV-12), an *Essex*-class aircraft carrier, had been christened and launched on August 30, 1943. After sea trials in the Chesapeake Bay and a shake down cruise near Bermuda, she docked at Norfolk, Virginia. On February 14, 1944, she sailed south to pass through the Panama Canal, and then on to San Diego, California. Carmen begins writing in his diary on February 27, 1944, the same day the *Hornet* arrives at San Diego.[1]

For the next year, Carmen, although it was against military

regulations, would record his thoughts and observations of life aboard ship. He would travel far from his Connecticut home. Far enough to have gone around the world four times. His first shipboard job was spotting planes on the flight deck. Then he became a mess cook, working below decks in the terrible heat.

All throughout his service aboard the *Hornet,* I found Carmen to be a cheerleader for his ship, air groups and pilots, his buddies, the naval task force, and the United States of America. Despite the worries and fears, Carmen and his fellow sailors did their parts to ensure our side's victory. We can be proud of the thousands of Carmens who served for us so that we can live free.

Carmen Franceschelli (at left) with his Sis, Barbara Franceschelli, and his Dad, Vincent Franceschelli.

'44 February

Saturday February 27, 1944

Docked at San Diego Calif, North Island Navy Air Base. Took off duds. Quite a few Marine Officers came aboard. Started to take a heavy load Jeeps & trucks. Had to respot the flight deck, all of our planes were secured on the flight deck. Liberty night. My first liberty since Christmas. Went to Balbo[a] Park, also San Diego Zoo.

Sunday February 28, 1944

*Worked on hanger deck, took planes aboard SB2C, TBF, F4F, F6F. Marines were working on the dock bringing supplies aboard. Worked late until both hanger and flight deck were loaded with planes, Jeeps, tractors, trucks, and supplies. Marines were getting ready to come aboard the U.S.S. **Hornet** to be transferred to Pearl Harbor.*

Monday February 29, 1944

*Left for Pearl Harbor with Marines about two thousand, also sailors rated and non rated. Altogether we had close to 6 thousand men on the U.S.S. **Hornet**. We left San Diego 0900. Marines were all trying to find a place where they could sleep. Most of them slept on the hanger & flight deck under planes. With so many men we were allowed only two meals a day.*

'44 March

Tuesday March 1, 1944

Ship crowded and very uncomfortable. Water hasn't been rationed so far. Getting ready to take Seamen 1 C test tomorrow. Water rough and plenty of Marines sea sick. No flight quarters can be operated. Have an escort of two destroyers. No work and taking it easy. Weather cold, but no doubt we will hit a much warmer climate soon. Had G.Q. [General Quarters] both morning & night.

Wednesday March 2, 1944

Plenty of Marines still sea sick, no work to do, and taking a nice rest. Marines were having a tough time. It wouldn't be to[o] bad if they didn't have to sleep on the hanger & flight deck. Its an awful mess. Some are sleeping on cat walks, in the elevator pits, mess hall & etc. Water a lot calmer today, also cool & a slight breeze.

Thursday March 3, 1944

Was on watch 1200 to1600 [hours]. Flight deck patrol. Lousy day windy & rain. Chow line plenty long, and still only two meals a day. Most of the Marines will be glad to get a shore and I don't blame them a bit. Pearl Harbor tomorrow. Water rough and a lousy cloudy day.

Friday March 4, 1944

Land was sighted 1300. Big beautiful mountain tops, and

boy they looked good. Few Marines and I went up to the flight deck so we could see what Pearl Harbor looked like while we pulled in to port. Docked about 1330. Plenty of work unloading planes, Jeeps, trucks, and all kinds of supplies. Marines had a tough time, and we all worked late until 0900 the next morning.

Saturday March 5, 1944

My first liberty night in Pearl Harbor. Went to Honolulu & Wakki Beach. Very beautiful country, beautiful palm trees, warm climate, and Wakki beach was a wonderful place. Along the shore it was shaded with big palm trees. During peace time it would be swell to be stationed there. Pearl Harbor was big, and well equiped with submarines, battle ships, carriers & etc. Ate plenty of ice cream, and drank alot of pineapple juice liberty night.

Sunday March 6, 1944

Went out to sea for new squadrons to practice landings on our carrier. Fighter Squadron #2 came aboard first. It was a hell of a mess, two crack ups. One was repaired, but the other was a dud. Quite a few flats [tires] today, and the fighter crew was green. Crew was green but the planes had bullet holes and one had five Japs to its credit.

Monday March 7, 1944

Flight Quarters SB2/c, TBFs & F6Fs came aboard. Had a big fire on the flight deck. One SB2/c burned to hell. Put out the fire, and brought what was left of the plane down on number 3 elevator. Elevator pit had about one foot of water. Had to use a pump to pump it out over the side. This was the first plane to burn aboard ship.

Tuesday March 8, 1944

Flight Quarters, squadrons were green. Worked on cleaning stations, and where getting ready to pull into port. Not much happened today.

Wednesday March 9, 1944

Pulled into port. Had G.Q. while entering. Took off duds. Held field day sweeping, scrubing bulkheads, and swabing decks. Band played some jive, and we had movies at night.

Thursday March 10, 1944

*Went out on the Harbor. Saw the old **Oklahoma,** which was sunk when Japs invaded Pearl Harbor. It was a hell of a mess covered with rust, and only a gun or two left. Went to show and saw **Slightly Dangerous.** They had a large canteen there, and boy was it good to get some good chow. Shore duty in Pearl Harbor looked good* *to me. The base was a beautiful one.*

Friday March 11, 1944

Worked all day taking supplies. Took enough food to last 4 months, and also take care of 1700 survi[v]ers if necessary. The **Hornet** *is getting ready for action, and we will be in battle very soon. At night a load of bombs were taken aboard.*

Saturday March 12, 1944

Out to sea. Left Pearl Harbor 0800. Planes came on. Had 5" firing run. Shot down two planes which were sent out to sea by radio control. Went to mass. Water nice and calm. Night firing run 20mm and 40mm.

Sunday March 13, 1944

Flight Quarters 0400. New squadrons came aboard. Baker planes were being equiped with 500 pound bombs. TBFs were equipped with dud torpeadoes. Sprinkler system turned on by accident. The hanger deck (aft) was flooded with water.

Monday March 14, 1944

Flight quarters 0315. Had early breakfast 0330. G.Q. was sounded 0630. Lasted about an hour. Pulled into Harbor 1600. First took off 4 duds. Loaded more bombs. Took twenty two more planes, and spoted them on the hanger deck. Worked most of the night taking amunition.

Tuesday March 15, 1944

This is a real day to remember. Left Pearl Harbor with a task force: U.S.S. **Hornet,** *nine destroyers, two heavy cruisers, & two converted carriers. Also same task force, and leading flag ship with us. The U.S.S.* **Bunker Hill** *with the same type of escort. Tonight they anounced over the speaker system that we are going to the Majuro Islands in the Marshalls. This is the real thing, and I'm sure we will see plenty of action soon.*

Wednesday March 16, 1944

Flight Quarters. Damp weather, little rain. Only about six planes patroling. Ships in regular formation and heading for the Marshall Islands. Expect to see action soon. Had mail call, and I received a picture of mother and me.

Thursday March 17, 1944

Flight Quarters. Not much work. Tropical rain storms, didn't last long. Had G.Q. tonight. Should be reaching Islands soon. No trouble so far.

Friday March 18, 1944

Lost two pilots. Two fighters colighted [collided] into each

other while in formation. Both pilots were killed. Plane from converted carrier made emergency landing on our carrier, no crack up, and he made a wonderful landing. The trouble—his tail hook wouldn't release.

Saturday March 19, 1944

Awful hot today. Worked on commissary stores. Didn't do much work. Supposed to reach islands some time tomorrow.

Sunday March 20, 1944

Land sighted, Majuro Islands. The islands were small except for one very large & wide one. They seemed to be circular, and not very wide, sand barges [bars] leading from island to island. All that you could see was that they were covered with palm trees. Anchored with rest of fleet. Was I surprised to see all the ships there, battle wagons, hospital ships, carriers etc. U.S.S. **Lexington, Bunker Hill,** *and our carrier were anchored close together. Must be getting ready for something big!*

Monday March 21, 1944

Still anchored at Majuro. Took alot of mines aboard today. I went up to the island [highest part of the carrier] to take a look at the Majuro Islands through the telescope. The Islands looked very beautiful with palm trees, and what a sight seeing the men who were stationed there swimming. Had the whole day off, and stayed on the flight deck to get a sunburn. Two admirals came aboard. It looks like we are getting ready for something big.

Tuesday March 22, 1944

Left Majuro 0800 with a large task force. Altogether we had 48 ships with us. The formation was very big, and that's all you could see was all kinds of ships. Had Flight Quarters. What a sight. Almost two hundred planes in the air. Tonight they announced over the speaker system that this was only one third of the complete task force, and also this was to be the biggest in Navy history.

Wednesday March 23, 1944

To this present time here are some of the ships we have with us. Battleships **New Jersey, Iowa, Alabama, Massachusetts, South Dakota, North Carolina.** *Cruisers* **New Orleans, Minneapolis, San Francisco, Indianapolis, Portland, Louisville, Chester, Pennsacola, Salt Lake City, Wichita.** *Carriers* **Lexington, Hornet, Bunker Hill, Cowpens, Monterey, Cabot.** *Destroyers* **Izard, Bradford, Cosdel, Brown, Conner, Morrison,**

Hickox, Caldwell, Halsey, Powell,
Bell, Burns, Choriette, Miller, The
Sullivans, Lewis Hancock, Porterfield
Hunt, Marshall, McDermont, Frazier,
Calloghan, Longshaw, Brancroft.

Thursday March 24, 1944

Passed the Equator today. That means I'm a shell back instead of a polly wog. Also we are a day ahead. By rights its Fri. but out here its Sat. I'm going by old time so I won't have to skip a [diary] page. Had flight quarters, and G.Q. everyday as usual. Skipped a day because we passed the National [International] Date Line. Still don't know our destination.

Friday March 25, 1944

Captain made a speach to crew & officers. He said the enemy we are going to meet very soon is murderous & etc. but we are better equipped & better men. He also told us that every man is to do his job in the best efforts he can. Had our ship refueled about noon time. Flight Quarters, crack up landing on flight deck. It was a fighter.

Saturday March 26, 1944

*SBD from carrier **Lexington** landed on our flight deck. Destroyer came along side for refueling. Tonight they* *announced our destination is going to be Palau Island. He said our task force is big enough to meet the Jap Fleet and he wished we will.*

Sunday March 27, 1944

This morning I saw in front of the **Lexington** *the carrier* **Yorktown.** *No doubt we will have many carriers with this task force. Had flight quarters most of the day. We already passed Truk & the Admiralty Islands. The water is very calm today, and this morning it looked beautiful to see so many ships sailing the calm sea with a beautiful sunset [sic] on the horizon.*

Monday March 28, 1944

Raid on Palau Island. G.Q 1300. Jap bomber was reported shot down. One of our fighters shot down a Jap bomber it was announced 1800. All night bombs were being loaded in planes. Tomorrow morning we will bomb the Island. Slept on G.Q. station all night long.

Tuesday March 29, 1944

This is it, planes took off to bomb Palau. All day long planes were loaded with bombs, and flew back to the island. T.B.F. blocked the harbor with mines. All ships there were sunk. Planes that came back had bullet holes in them, some had big parts torn out. G.Q. alarm sounded, nine Jap bombers were shot down. I could see smoke from a distance. The Navy air force is bombing the hell out of the island. Slept on G.Q. station. At night Jap bomber was shot down close by ship.

Wednesday March 30, 1944

Planes took off again to bomb island. Second bomber squadron came back with bombs. I guess they must have bombed the hell out of Palau. Sunk quite a few ships, but the darn Jap fleet skiped again. Since the beginning of battle I slept on the hanger deck, and ate sandwiches, which were sent up by bomb elevators from the galley. All hatches are closed, and we can't wash or put on clean clothes. All night long planes were being loaded with bombs, torpeadoes, and amunition. Slept on hanger deck again.

Thursday March 31, 1944

Planes took off again. This morning two Jap cargo ships were sunk. I saw smoke in the distance. We lost quite a few planes, but most were crack ups. About three or four planes were shot down, and many came back with bullet holes. Remember one gunner lost two fingers from a bullet wound. At last set condition yoke was sounded. We had a hot supper, and was it good to take a shower & to sleep on a matress for a change.

'44 April

Sunday April 1, 1944

Today is Palm Sunday. This morning they buried two men at sea. Finished bombing Island, and are heading back to Majuro. Complet[e] task force bombed two Islands Palau & Yap. Water nice and calm today. Hope next Palm Sunday I'll be home.

Monday April 2, 1944

Lousy day. Not much happened. We worked on our cleaning station. Had Flight Quarters. Found out that we lost eight planes. Two shot down, four missing, and about three are duds.

Tuesday April 3, 1944

Weather still lousy, its raining. The ship is a mess, and most of the crew are on cleaning stations. Band played tonight, and was good to hear some jive.

Wednesday April 4, 1944

Lousy out, its still raining. Not much to do so I studied my course book today. Since we are going back to Majuro, I'll be receiving mail, and boy will it be good to hear from home. Even though its raining its awful hot.

Thursday April 5, 1944

Tonight it stoped raining & its starting to clear up. Hope it will be a nice day tomorrow. Went back to regular time. No day ahead. Didn't do much work. Should reach Majuro soon.

Thursday April 6, 1944

Reached Majuro 0900. It started to rain again today. Ships anchored close to island. We had a movie tonight and the band played. Saw in Hornets News Flash today [that] every ship that had not fled from Palau, Yap, and [blank] harbors was sunk or damaged. Three Japanese warships were caught & sunk. Fourteen merchantmen, two tankers, and twelve cargo ships were sunk by submarines in separate actions. 27 American planes lost & no ships were damaged or lost.

Friday April 7, 1944

Today is Good Friday, & soon Easter will be here. At last mail call, and boy was it good to hear from home. Worked most of the day taking supplies aboard. Had another movie tonight.

Saturday April 8, 1944

We took a seaplane aboard which we will use while here at Majuro. Had a Happy Hour tonight, and an officer played Artie Shaw's "Concerto for Clarinet." Tomorrow will be my first Easter away from home. I'm sure going to miss mom's super holiday specials.

Sunday April 9, 1944

Went to High Mass and received Holy Communion. Worked on our cleaning station, and wishing I was home today. Saw the Admiral and helped to lower the seaplane which he was in.

Monday April 10, 1944

Just working on cleaning station for the last few days. Was on work party taking supplies aboard. A Seabee from Majuro told me that eight thousand men were stationed on the island.

Tuesday April 11, 1944

Was on work party. Took a fighter aboard which landed on the **Bunker Hill** *when we were at Palau. Still taking it easy. The band played and we had movies tonight.*

Wednesday April 12, 1944

Worked on our cleaning station. Respot planes on hanger & flight deck. Had a movie tonight. Tomorrow we leave Majuro Island. Do not know where our next atack will be.

Thursday April 13, 1944

Left Majuro 0930. Respot planes, both hanger & flight

deck. Took a dud down from the hanger deck. Worked on our cleaning station.

Friday April 14, 1944

Worked on our cleaning station. Had Flight Quarters, crack up on the flight deck, landing gear colapsed while fighter came in for a landing. Still don't know where we are going. Cruiser was shooting at skee slead [sled] we were towing. Those shells were bursting darn close to our ship. To me it look[ed] like they were doing very good shooting.

Saturday April 15, 1944

Flight Quarters. G.Q. Planes were practicing diving & shooting at a skee slead. Fighters and bombers also droped water bombs. I could see the water bombs falling through the sky. Tonight they announced over the speaker that we are going to New Guinea. Fighter lost control of plane in air, and crashed into the sea. Pilot got out and a destroyer picked him up.

Sunday April 16, 1944

A day ahead. Had flight quarters. Worked on cleaning stations most of day. Water calm. Should reach New Guinea soon.

Monday April 17, 1944

Flight Quarters. Still on cleaning stations. Lousy day today. Destroyer pulled alongside to deliver official mail.

Tuesday April 18, 1944

Had flight quarters 0300. 4 night fighters made first hop/ Radar picked something up. G.Q. sounded about 0400. Think it was Jap bombers, but anyway nothing happened. Had a very close call today, two tankers almost smashed into our ship as we made a sharp turn, just missed our stern by inches. Ate at battle stations and stayed on flight quarters stations all day. ~~Reach New Guinea tomorrow, and will attack raid most important airfields also troops will land.~~ [Crossed out by Carmen.]

Wednesday April 19, 1944

Had flight quarters all day. Worked on cleaning stations. Refueled about four destroyers. Reach New Guinea tomorrow. Task force will bomb most important air fields, shell Island and land troops.

Thursday April 20, 1944

Got up 0550. 1st raid 0700. Ran into heavy rain storm but it didn't last long. Two SB2C went over the side while take off. Night fighters bombed & strafed 13 Jap planes on Island, also blew up ammunition dumps. SB2C came back with bullet holes. It was a mess, looked like it had a dog fight. Close enough to see island of New Guinea 0900, saw palm trees in water. SB2C made hell of a crack up on the flight deck. We pushed it over the side after taking out usefull parts.

Friday April 21, 1944

One of our SB2C shot down a Jap Betty. Had flight quarters. More raids on the island. Saw troops on barges head for island 1500. What we did to New Guinea in two days: damaged on island twin engine planes, 4 single engines, 67 planes destroyed, 20 twin engine bombers and 15 single, wrecked air field Alondi fuel & ammunition dumps & supplies, sunk one heavy freighter. Our ship made 750 air hours & droped over 100 tons of bombs.

Saturday April 22, 1944

Our patrol squadron at 1145 shot down a Jap Betty, took one Jap survivor aboard. He was picked up by destroyer.

Sunday April 23, 1944

Shot down a Betty very close to ship, could see smoke in the distance. Made more raids. Went into G.Q. Picked up something on Radar but nothing happened.

Monday April 24, 1944

A converted carrier in our task force shot down Jap Betty, destroyer picked up three Jap survivors. Saw pictures our planes took while bombing New Guinea airports A/A and Sawar. What a mess, could see Jap Bettys wrecked on field. Altogether our planes destroyed 119 planes.

Tuesday April 25, 1944

Night fighter cracked up late last night. Five of our pilots got credit for shooting five Jap planes in the air.

Wednesday April 26, 1944

Passed Admiralty Islands. Took eight new Bakers, two new fighters from island (could see Island) for replacements.

Thursday April 27, 1944

Mail call. It was darn good to receive letters from home. Sis tells me Jessie was on Truk raid. We make our first strike on Truk April 29. Passed [Inter]national Date Line. No more day ahead.

Friday April 28, 1944

Passed another Admiralty Island. Had mail call. Received a letter from Jessie. Mom's package finally reached me. It was a mess, after waiting for it so long.

Saturday April 29, 1944

*Made a raid on Truk first flight 0600. Made 7 raids today. Two crack ups on flight deck, TBF & F6F. One of our TBFs was shot down. Very high wind force of propeller blew a man off the flight deck. Destroyer picked him up. TBF came in with its elevator & rudder messed up from anti aircraft fire. **Bunker Hill** while shooting at six Jap planes hit one of our own cruisers with a five inch shell. It killed five men.*

Sunday April 30, 1944

*Made three flights to Truk. We had three crack ups, two SB2C, one fighter. Had to dump one SB2C over the side. **Cowpens'** TBF with half of his stabilizer shot off made a beautiful emergency landing on our carrier. His radio gunner was found dead. G.Q. tonight. Part of task force was shooting at two Jap snipers. I could see them firing their guns. SB2C caught fire while made crash landing. Our fire fighters put out the fire.*

'44 May

Monday May 1, 1944

Made a raid on one of Truk Islands. Three planes were wrecked while landing, two SB2Cs & one fighter. Today we had the worst accident that ever happened on our carrier. An SB2C came in for an emergency landing. By accident a hundred pound bomb droped from the SB2C rack. It exploded and killed two men, one a good friend of mine, and injured six men. Battle wagons, cruisers & destroyers shelled the Island at 1500.

Tuesday May 2, 1944

Had a hell of a time. Inniciation [initiation] from a pollywog to a shellback will be given tomorrow. Today we had a free for all and boy what a fun [time] we had. Elmer & Moose started a fight on the flight deck with water hoses. Hornets New Flashes attack against Truk, Satawan, and Panope lasted three days. War ships shelled Japanese installations, while planes bombed and gunned targets of all sorts. None of ships damaged & aircraft loses were light, approximately 30 flight personnel are missing. Over 60 enemy planes were shot down in aerial combat over Truk & an equal number were destroyed on ground. Droped 800 tons of bombs on Island.

Wednesday, May 3, 1944

Today I became a shellback and what a beauty of an inniciation. We had to go through the middle, dumped in a tub of salt water, and what a haircut they gave me. I'm practically bald. Today I start my three months of mess cook. U.S.S. Hornet Late News Flashes: ~~Powerful Naval task force under command of Vice Admiral Mitsher.~~ *The attack against Truk, Satawan, and Panope lasted three days. War ships shelled installations, while planes bombed and gunned targets of all sorts. 60 planes shot down in aerial combat, destruction of 126 Japanese planes & sinking of 17 enemy warships were reported yesterday.*

Thursday May 4, 1944

Anchored at Kwajaleen atoll in the Marshall Islands. We are only about 250 miles away from Majuro. We anchored 0800. Had a mail call, and I received a letter from home. They showed a movie at night. Late News Flashes: destruction of 126 Japanese planes, sinking of 17 enemy warships and the damaging blow ever struck at Nippon's Central Pacific fortress of Truk were reported yesterday. Truk was blasted by record load of 800 tons of explosives, in coordination blows Truks primary eastern out post Panope & Patawan Islands ever shelled by big naval guns

& blasted by carrier live bombers & torpedoe planes. Our losses light. 30 airmen were missing.

Friday May 5, 1944

Work parties, food supplies, bombs, ammunition. We are taking an awful lot of supplies aboard. Marine pilots out on Island fly Corsair fighters. Boy are they fast. Somehow this island looks a lot like Majuro. Guess when you see one you see them all.

Saturday May 6, 1944

Still taking supplies. Today we had trouble. They searched all lockers because of two revolvers being stolen. Received alot of old papers from home, they had plenty of news about our raids.

Sunday May 7, 1944

Still work parties and working on cleaning stations. We have two seaplanes on hanger deck for officer pilots to use while anchored here. So far we always been having movies while anchored. Hornets News Flashes New Guinea: heavy bombers struck Wewak airdromes, bivouacs, and supply areas with 105 tons of bombs. Airdromes were damaged, and bridges, gun emplacements, & buildings destroyed, many large fires were started. Night patrol in the area of successive nights destroyed four barges, damaged seven others, and silenced shore batteries, while air patrol harassed communication and installations in the Spik Valley.

Monday May 8, 1944

Liberty parties. Eight men out of each section are going on the beach, just to get off the ship for a change. These island[s] are swell places to go swimming. I won't be able to get off the ship since I'm on mess cooking. It sure has been a long time since I've been on land.

Tuesday May 9, 1944

Still having liberty parties. Also games on hanger deck & flight deck such as basket ball, tennis, volley ball & etc.

Wednesday May 10, 1944

Not much to say. We will be hanging around a long time before we make our next raid.

Thursday May 11, 1944

Still at Kwajaleen Island. Leaving for Majuro tomorrow. Nothing of interest happened.

Friday May 12, 1944

Left Kwajaleen Island 0900. Sure wish I had a chance to go on one of those liberty parties. Darn my luck. It would be interesting to see one of those Jap islands, I mean really walk around one & get to see what it looks like.

Saturday May 13, 1944

Reached Majuro Island 0900. Most of crew working on work parties or on cleaning stations. Nothing of any interest happened.

Sunday May 14, 1944

Mothers Day went to mass and received Holy Communion for mom. Had a good dinner for a change, I mean for aboard ship. Boy do I miss my mom's home cooking.

Monday May 15, 1944

Tonight we had a movie. A CO2 bottle went off. We all got scared. One fellow was pushed over the side by accident and he drowned.

Tuesday May 16, 1944

Ship is doing an awful lot of practicing with anti aircraft & 5 inch guns. Today they did alot of practicing.

Wednesday May 17, 1944

Still hanging around. Nothing of importance.

Thursday May 18, 1944

*Mess cook keeping me busy. Hanging around, nothing of importance. Alot of war ships anchored here. I can see 5 **Essex** class carriers not counting ours.*

Friday May 19, 1944

Most of the ships in the past few days have been having firing runs. Still on mess cook & its awful hot below decks. Holding field day today. Using hanger deck for a gym playing basket ball, volley ball & etc. Had a movie & the band played.

Saturday May 20, 1944

There are 5 big carriers along side of us. We have all kinds of fighting ships anchored here near.

Sunday May 21, 1944

Captain Browning will be leaving us soon & a new Captain will take his place. Holding field day & having firing runs.

Monday May 22, 1944

Today I noticed we have a lot of troop ships. We must be getting ready to invade some Jap island. Holding field day. Cruisers had firing runs. Had a movie tonight.

Tuesday May 23, 1944

Still in port. Holding field day, firing runs, still using hanger deck for gym. Took a lot of supplies aboard in the past few days. Had a movie.

Tuesday May 24, 1944

Holding field day, had firing runs. Had a movie. Nothing of interest, still hanging around.

Wednesday May 25, 1944

Holding field day. Firing runs 40mm & 20mm. Still taking supplies aboard. We are getting ready for something big. Band played & we had a movie.

Thursday May 26, 1944

Holding field day & had firing runs. Nothing of any importance.

Friday May 27, 1944

We should be leaving soon. Its been awful long time we have been hanging around Majuro Island.

Saturday May 28, 1944

Holding field day. Getting ship cleaned up for Admiral Inspection. Had firing runs. The band played and we had a movie.

Sunday May 29, 1944

Our Captain Browning left ship today, and Captain _____ [blank space in diary] took over. Browning made a speech before he left. Said this was the first carrier to be in enemy waters within four months. Had dress parade and new Captain made speech, also Browning before leaving.

Monday May 30, 1944

Worked all day for Admiral's Inspection. Admiral and Captain _____ [blank space in diary] will make inspection tomorrow. In this past month nothing of interest except we were getting ready for invasion of some Jap island, which we found out to be Saipan.

Tuesday May 31, 1944

Had our inspection. Had to wear whites. Inspection was a success, and the ship was a lot cleaner.

'44 June

Wednesday June 1, 1944

Happy hour. **South Dakota** *entertainers besides ours. Art's song "Mood Majuro" may be published.*

Thursday June 2, 1944

Left Majuro 0635. We are going out to sea for firing runs. Had G.Q. while leaving Majuro. Flight Quarters was sounded 1300.

Friday June 3, 1944

Anchored Majuro 1600, had firing runs while anchored and shot down a good number of sleeve's. Today is my birthday and I'm twenty years old. Sure wish I could be home ~~today~~.

Saturday/Sunday June 4, 1944

Went to Mass. Had firing runs, 5", 40mm & 20mm every day while being in port.

Sunday/Monday June 5, 1944

G.Q. Picked up something. Don't know what it was but we had G.Q. twice.

Monday/Tuesday June 6, 1944

Left Majuro 1300 & this time we are going Jap hunting. G.Q. while leaving the Island. Lousy day, rain & cloudy.

Tonight was announced the Invasion of Europe was started.

Tuesday/Wednesday June 7, 1944

Flight quarters. Aerial combat practice. Firing runs. Two TBF crack up on flight deck.

Wednesday/Thursday June 8, 1944

Flight quarters. Aerial combat practice & also firing runs 20mm, 40mm, 5".

Friday/Saturday June 9, 1944

Flight quarters most of day. Also had firing runs. Should reach our destination soon.

Saturday/Sunday June 10, 1944

Our Captain announced we strike Guam. Rest of task force will strike Saipan and Tinian tomorrow. Also said a B24 just reported shooting down an Emily this morning. The way Captain talks we are in for alot of trouble. These three islands have big Jap air fields. He also said you can tell your grand children about this raid if you live to tell it.

Sunday/Monday June 11, 1944

Bataan *fighters shot down Emily down early this morning. Also* ***Yorktown*** *fighters shot one down.*

Captain made speech. Said that part of task force radioed Guam was ready for us. They know we are coming. Admiral radioed back we will make every bomb & bullet count, and down with the yellow bastards. Our fighters shot an Emily down 1435. Seventeen of our fighters shot down 25 Jap planes out of the sky, and we lost one fighter. He made a good water landing, and may be picked up. Took two Jap prisoners aboard. One was named Mosso Goto.

Monday/Tuesday June 12, 1944

First flight took off 0630. Saw one SB2C go in the drink, both pilot & gunner were picked up. Captain announced **Bataan** *fighters shot down 2 Jap planes down early this morning. For the day our task force fighter planes shot down 41 Jap planes out of the sky & destroyed probably 5 on the ground. Our Captain announced our fighters shot down eleven planes out of sky. Other part of task force Saipan Tinian shot down 84 planes for the day. We lost t[w]o pilots, a bomber & fighter pilot, bomber in drink, fighter shot down over Guam.*

Tuesday/Wednesday June 13, 1944

Made seven raids on Island. We were close enough to see Guam 1730. Our planes damaged a Jap cruiser & a Jap destroyer. Don't know if they sunk, but direct hits were made. T[w]o F6Fs duds damaged from pulling out of dives & duds riping stabilizer to bits. Quite a few planes damaged from anti aircraft fire. Fighter pilot was picked up by a sub.

Wednesday/Thursday June 14, 1944

Ships shelled Guam, Tinian & Saipan. Troops landed on Saipan. No bombing raids today. Had to refuel destroyers. Heading for Bonin Islands full speed ahead, after the Jap fleet which was reported sighted there.

Thursday/Friday June 15, 1944

Only 500 miles from Japan. Started raids on Bonin Island Chichi Jima sea plane base. Sunk a cargo ship. Fighters shot down 17 Jap planes, lost two planes SB2C [and] F6F. Crack up on flight deck TBF. Destroyer sunk a Jap troop ship and close to two hundred Jap survivors were ~~taken~~ rescued. Planes had a hell of a time landing. Looks like Jap fleet fled again.

Friday/Saturday June 16, 1944

This morning we took Jap prisoners aboard, 123 Jap survivors which the destroyer picked up yesterday. I watched them transfer the Japs to our ship. They are using ship service number one for a prison. Our planes went on search parties as close as 250 miles from Japan.

Saturday/Sunday June 17, 1944

Planes went up on search parties looking for the Jap fleet, but no luck so far. All our planes are loaded & plenty of bombs & ammunition. Just laying around waiting for the Jap fleet. We can get Tokyo broadcast over our radio. Tokyo Rose seems to get everything the opposite, and tells the Japanese people every thing backwards. The Jap prisoners are being well guarded by our Marines. Planes made a patrol for five hours looking for Jap, altogether the planes patrolled the sky for ten hours looking for the Jap fleet.

Sunday/Monday June 18, 1944

Went back to Guam & made another raid. Our fighters shot down 52 Jap planes down. Our raids destroyed 223 planes, task force shot 108 out of sky, & destroyed 115 on ground. A fighter came back with plenty of bullet holes, a couple right through the green house, another carrier. The Jap fleet was sighted by Tinian. These planes we shot down are planes which took off Jap carriers and land on Guam. That means some of the Jap carriers are not equiped with planes. Could see ships in task force shoot down a few Jap planes. Jap fleet is believed to be close by.

Monday/Tuesday June 19, 1944

At last the Jap fleet was sighted. Our planes left carrier

1610 to hit the Jap fleet. Our planes sunk a carrier as big as ours. Carrier had very few planes because [some] of them landed on Guam, but we found out by shooting 335 planes out of the sky (that is including all of task force). Sunk one carrier & probably more ships. Quite a few damaged. Lost alot of our planes on carriers too far from fleet (300 miles) & planes no gas droped in the drink. Admiral & Captain had every ship in the task force light up all lights on ship. Admiral said he would light up the whole South Pacific [as] long as we get our boys back safe.

Tuesday/Wednesday June 20, 1944

Many pilots were being picked up after what happened last night. A flight took off to face Japs fleet again, but they fled. Pilots would see an awful lot of oil boxes & junk in water. So far we are sure one big Jap carrier was sunk.

Wednesday/Thursday June 21, 1944

Still [picking] up survivors. No trace of Jap fleet. One pilot who was picked up near where the Jap fleet was hit said he actually saw the Jap carrier blow up & go to the bottom. Also one was very badly damaged, but it vanished out of sight. Its believed this carrier may have sunk to[o]. Hornet's News Flashes: Tokyo Radio last night broadcast the following essurtion [assertion] of

Imperial Headquarters. These American ships were sunk today in Marianas 1 battleship, 2 heavy cruisers, 1 sub, 1 destroyer, damaged 2 battlewagons, 4 CV type carriers, 4 heavy cruisers, 6 transports, & 1 unidentified ship. More than 300 American planes were shot down. (No mentioning Japanese losses.)

Thursday/Friday June 22, 1944

Getting ready to strike the Pagan Island. Had a funeral service for one pilot who was killed in a water landing. Refueled ships, band played and it was good to hear some music. U.S.S. Hornets Late News Flash: June 18, 353 enemy aircraft were shot down, of which 335 were destroyed by our carrier aircraft and 18 by anti aircraft fire. June 19, one Zuikaku carrier three one-thousand pound hits believed to sunk,1 Hayataka carrier was sunk, one Hayataka severely damaged, one light carrier received at least one bomb hit, one Kongo battleship was damaged, one cruiser was damaged, three destroyers damaged, one of which is believed sunk. Three oilers were sunk, two tankers severely damaged & 15 to 20 defending aircraft were shot down. Our los[s]es 49 aircraft, including many water landings.

Friday/Saturday June 23, 1944

*Made strike on Pagan Islands. Our **Hornet** planes shot*

down 4 Zeros, 1 Zeke & a Betty. Had an SB2C crack up on flight deck, pilot and gunner were O.K. A pilot was picked up who was in the water for eleven days. Four small cargo ships and one sampan sunk, two small cargo ships, & twelve sampans damaged, four enemy aircraft destroyed, and two probably destroyed on the ground. A flight consisting of one twin engine bomber and five zeroes was intercepted some distance from our carrier force, was shot down. A wharf & fuel dump were destroyed, and buildings & run ways were damaged. Lost one F6F fighter.

Saturday/Sunday June 24, 1944

Bombed Bonin Islands, our fighter squadron is superior. They broke the navy's record for shooting down the highest number of enemy planes down in one raid, shot down 67 Jap planes out of the sky. Had G.Q. 1030 to 1320. Saw the **Bataan** *open fire on a Jap plane. This mornings report were as follows:* **Yorktown** *19, the* **Bataan** *13,* **Hornet** *33. Saw a fighter make a water landing, pilot was picked up & he is O.K. Had a movie tonight. Was just about to hit the sack when G.Q. was sounded, radar picked up Bogies about fourteen miles away from task force.*

Sunday/Monday June 25, 1944

Went to Mass this morning. A patrol squadron went up 1130. Heading for Eniwetok in Marshalls for anchorage to get new replacements & supplies. Had a movie tonight. U.S.S. Hornet Flashes: a Pacific fleet sub torpeadoed a Shokuku class carrier on 18 June. Three torpeadoe hits, carrier probably sunk. Our strike on Japanese fleet June 19 one small carrier reported damaged or sunk, two Jap twin engined bombers were shot down by carrier aircraft returning to our carriers.

Monday/Tuesday June 26, 1944

Should reach port tomorrow. Had flight quarters 1130 to 1600. SB2C dud was pushed over the side number two elevator. Captain made a speech about our work during these last raids, he said God was with us and with his help we are bound to win. Our Captain is a swell guy, and our ship is sure getting ahead fast. At this time we hold the worlds carrier record for shooting enemy aircraft out of the sky, and we only [have] been out to sea for about 5 months. Some carriers have been out here close to two years, and we have a better record then [than] them.

Tuesday/Wednesday June 27, 1944

Anchored at Eniwetok. Anchored next to the U.S.S. **Franklin** *which just got out of the States. Many ships*

anchored here, especially troop ships. Started to take supplies aboard. We won't stay in port long. Had firing runs 20mm & 40mm. Hornets News Flashes: carrier task force swept Iwo Jima Island on 24 June. Sixty or more enemy aircraft of a force which attempted to intercept our fighters were shot down. Twelve of the enemy planes found our carriers, all were shot down by our combat Air Patrols. We lost four fighters.

Wednesday/Thursday June 28, 1944

Still anchored at island. Taking supplies & ammunition aboard. The mess hall was filled with bombs 2000 pounds, 1000, 500, & etc. We should be leaving the island very soon. Had a movie tonight.

Thursday/Friday June 29, 1944

Still anchored at Eniwetok. Jap prisoners were taken off the ship today. Still taking supplies aboard. Nothing of interest happened.

Friday/Saturday June 30, 1944

Left Eniwetok, saw troop ships form a line, and what a line! Easy over thirty ships. Took replacement planes aboard, new fighters and new SB2C. Left island 1300. The invasion of Guam should [be] taking place soon.

'44 July

Saturday/Sunday July 1, 1944

We had flight quarters & firing runs. Took my final progress test for Aviation Machinist Mate. One more test to go.

Sunday/Monday July 2, 1944

Sunday went to mass. Had flight quarters, F6F replacement fighter went into catwalk while taking off. Pilot was O.K.

Monday/Tuesday July 3, 1944

Fighter Squadron took off 1300. Came back 1815 with 33 Jap planes to their record & 3 probably, 12 near Bonin Island. Tomorrow we make another raid on the Bonin's Chichi Jima & Iwo Jima.

Tuesday/Wednesday July 4, 1944

Two of our night fighters shot down 7 Jap planes. Both pilots landed on **Yorktown.** *One pilot had his collar bone broken from flying scrap metal, he is going to be O.K. 1st strike on Bonins 0645. For the day 24 cargo ships were badly damaged, one mine layer was sunk, 1 cruiser was badly damaged. Bonins used for seaplane bases. They were damaged very badly by our planes. Ships at noon time went in and shelled the Island. One SB2C shot down, pilot and gunner's first hop. Pilot bailed* *out but enemy anti aircraft killed him on the way down. Search planes went as close as 160 miles from Japan.*

Wednesday/Thursday July 5, 1944

Held field day on cleaning stations. Left Bonins & heading for Guam. One plane which was very badly damaged by anti aircraft fire was shot off the catapult into the drink. Had G.Q. tonight, radar picked up Bogies 20 miles from ship. Night fighters were ready to take off but they didn't.

Thursday/Friday July 6, 1944

Made one strike on Guam, not much anti aircraft fire, & little enemy resistance. Tonight two of our SB2C went on search hop for missing pilots. They sighted two pilots, one 5 miles from shores of Guam & the other one fourteen miles away, droped liferafts, radioed their positions & both pilots were picked up. Battle in Saipan a very bloody one, we lost a lot of Marines but the Japs lost about five times as much as we did. Last reports over 2000 Marines killed, Japs over 9000 killed.

Friday/Saturday July 7, 1944

Last night two of our night fighters went up 2115 & shot down 2 Jap Betties. You could see one of our night fighters sho[o]t down a Betty 20 miles from ship, Betty burst into flames & hit the water. Make another strike of Guam tomorrow.

Saturday/Sunday July 8, 1944

*Flight quarters, 2 patrol flights & one strike on Guam.
A destroyer picked up 5 Jap survivors which were from
a Betty which our pilots shot down yesterday. I saw
them when they took them to the brig. Two of our TBFs
landed on Saipan. They came back with a lot of silvoners
[souvenirs] the Marines gave them. Late News Flashes:
Pagan Islands was attacked by carrier aircraft on 4 July.
The runway at the airfield and adjacent buildings were
bombed & strafed. Barracks & supply facilities at Guam
Island were bombed by carrier aircraft on July 4, starting
large fires. We lost one SB2C (Conquest of Saipan).*

Sunday/Monday July 9, 1944

*Late News Flashes: reports from a fast carrier task force
Group which attacked Chichi Jima in the Bonin Islands
on 3 July & also attacked Haka Jima. A group of several
enemy ships located eighty miles northwest of Chichi Jima
was attacked resulting in the sinking of two destroyer
escort type vessels & damage to a medium cargo ship. At
Chichi Jima following results, one small oiler, one medium
ammunition ship & one medium cargo ship & a minelayer
& one destroyer damaged, several were beached. Haka
Jima two small cargo ships & minelayers damaged. We*

lost five pilots, four gunners & seven planes missing.
Made a strike on Guam today. Cruiser and battlewagons,
destroyers shelled the island.

Monday/Tuesday July 10, 1944

Saipan surrendered and is in well in hand. It looks like
we will invade Guam soon. Two of our TBFs landed
on Saipan, & they came back with a lot of silvoners
[souvenirs]. We are supposed to stay out around Guam
until July 25, just making single strikes a day until they
are ready to invade Guam. Two planes are going to land
on Saipan tomorrow again & some of the mess cooks and
I got some apples for one of the gunners who is going over
to Saipan to give to the Marines there. Two [?]

Tuesday/Wednesday July 11, 1944

Made a strike on Guam. A destroyer picked up a 1C Navy
radio man who signaled from Guam with a mirror to a
destroyer. They got his signal & told him to swim out to
destroyer. He did & was picked up & was brought [to]
our carrier. What a story he had to tell. He was on Guam
[in] 1941 when the Japs invaded Guam & took the island
over. He fled to the hills with five other men. There the
natives took care of him for 2½ years. Four of [the] men
were killed by Japs in 1942 & one disappeared one day.
The Japs got him no doubt. Some of our planes landed

on Saipan, talked to Marines and saw thousands of dead Japs.

Wednesday/Thursday July 12, 1944

Two patrol flights. We had a movie at night. The Captain made a speech about what our air group did in the last 30 days. It goes something like this, destroyed 223 Jap planes & 23 probablies. Damaged close to 50 cargoe ships at Bonin Island & Guam. Sunk Jap carrier while hiting fleet. We made 7 direct hits. Our ship should receive credit for sinking carrier. We lost 14 men in our air group, 18 Bakers, 3 TBFs & 4 fighters. The record before our air group broke it was 60 for a day raid, our fighters shot down 67 & besides we had a movie that night.

Thursday/Friday July 13, 1944

Had two patrol flights. Running short on flour & butter. Gunner I know landed on Saipan & he showed me a Jap shord [sword] a marine gave him for a silvoner [souvenir]. It was a beauty with gold trimmings. One drink [plane] went in the drink on the way back from Saipan, all crew got out & were picked up by destroyer.

Friday/Saturday July 14, 1944

Made a strike on Rota & Guam. We could see Guam this morning. We were about twenty miles away. Had two

patrol flights. We are going to keep on hiting Guam until they invade July 21. Guam was shelled today.

Saturday/Sunday July 15, 1944

Two patrol flights. A strike on Guam.

Sunday/Monday July 16, 1944

Flight Quarters, two patrol flights & one strike on Guam. Had a movie. Late Hornet News Flash: read about the Barnum & Bailey [Circus] fire in Hartford. It is now clear that Saipan Island was built up by the Japanese as the principal fortress guarding the southern approaches to Japan, and as a major supply base for Japans temporary holdings in the South Sea Area. It was learned on Saipan that on July 7, Vice Admiral Chieichi Nagumo, Commander in Chief of the central Pacific for the Imperial Jap navy, was one who met their death on Saipan. He was the one who was in command which attacked Pearl Harbor Dec. 7, 1941.

Monday/Tuesday July 17, 1944

TBF went into the drink this morning. Young Conary was a member of its crew. He was a friend of mine & also the one who gave me a silvoner [souvenir] from Saipan, part of a Jap cardboard ammunition box with Jap writing. Anyway all of crew was saved & picked up by the destroyer. Conary's pilot made a beautiful water landing, we could see the plane hit the drink from the flight deck. Made a strike on Guam, also two patrol flights.

Tuesday/Wednesday July 18, 1944

Made a strike on Rota. Could see Rota Island most of the day, we were only eleven miles away. Beano flew over Saipan, & told me the place was wrecked, nothing left of the town, also the town in Guam is in the same condition. Made two strikes & had one patrol flight. Our battle wagons, cruisers & destroyers have been shelling Guam & Rota very frequently. Three crack ups on flight deck today, 2 SB2Cs, one TBF.

Wednesday/Thursday July 19, 1944

This is it. Made four strikes on Guam hiting it very hard. Planes went in low strafing & also a very heavy load of bombs were droped. Hornet Late News Flash: Guam Island was shelled by units of the Pacific Fleet, and bombed by carrier aircraft on July 14. Gun emplacements and the airfield at Orate were principal targets. Four enemy aircraft were destroyed on the ground. Hornet News Flash: battleships, cruisers, & destroyers on the 15th heavily shelled Guam gun emplacements and other installations on the Island. The following day aircraft carrier force attacked Rota Island.

Thursday/Friday July 20, 1944

Made six strikes on Guam. Tomorrow we invade Guam. Our ships shelled the island and our planes were radioed where to come in & bomb certain parts of Guam. Late Hornet News Flash Pearl Harbor: Pacific Fleet Commander Admiral Nimitz announced today that powerful American surface units on Sunday continued their pre invasion shelling of Guam. The Pearl Harbor war bulletin said that powerful armada included an undisclosed number of battle ships, cruisers & destroyers. A similar attack was reported Monday night. At the same time American planes operating from Saipan assisted in the neutralization of enemy base Tinian.

Friday/Saturday July 21, 1944

We received mail which came off of Saipan. Made invasion today about seventy troop ships anchored away from Guam shores. Barges were lowered & Marines hit the beaches of Guam, with little enemy resistance so far. Heard over the speaker reports from our Navy planes to Congo head of Marine forces invading Guam. Reports were made in code & our planes had the code word Whirlwind which meant **Hornet's** *planes reporting. For the day our troops did very well & met with little enemy resistance, but they expect to face the enemy soon.*

Our planes were flying over Guam all day hitting enemy targets.

Saturday/Sunday July 22, 1944

Anchored at Saipan & Tinian for ammunition, what a sight. The island was quite a big one. We could see our destroyers still shelling Tinian. Today I saw more fat flies then I ever saw in my whole life, the cause was because of the thousands of dead Japs & Marines. A couple of Marines and Seabees came aboard our carrier & they sold Jap pistols & Jap money. Saipan must have been quite a sight. One of the Coxswains I know who pilots the Admiral's barge told me of the thousands of dead Japs they have piled up in stacks on the island. We left Saipan 1700, and are on our way again.

Sunday/Monday July 23, 1944

Had flight quarters, patroling squadron. A gunner told me today that we are heading for Palau, Yap, & then back to the Bonin Islands only 600 miles from Japan.

Monday/Tuesday July 24, 1944

We were suppose to make a [raid] on Woleai, but it was cancelled. Had two patrol flights. Getting very short on food especially flour. During meals two slices of bread per man & at times only one per man. Hornet News Flashes:

good beachheads have been secured on Guam Island by Marines & Army troops. The troops advancing inland are meeting increasing resistance in some sectors. On the nineteen July six hundred & twenty tons of bombs & one four seven rockets were expended in attacks on Guam by carrier aircraft. Naval gun fire and aerial bombing were employed in support of the assault troops up to the moment of landing.

Tuesday/Wednesday July 25, 1944

Received Captain's speech telling us that we did a good job under his command & also the trouble our planes had coming back after hitting the Jap fleet. Patrol flight took a picture of Ulethi Island, a very small island near Yap in the Pacific. Made a strike on Yap Island. Not much enemy opposition aircraft. Planes bombed their new airfield on Yap with little trouble from Japs. We had a G.Q. sounded 1315. I went up to the hanger deck. Our planes were already up in the sky making a search & patrol over our ships. G.Q. lasted 1430, radar picked up bogies in all different directions.

Wednesday/Thursday July 26, 1944

Made a strike on Yap, little enemy opposition. One of our TBFs came back with a large shell hole made by a 40mm shell which exploded in the wing. Pilot went in too low for a picture & that the reason why they were hit. Still running short on food. We are going to have some big working parties once we anchor at some Island for supplies. Still on mess cook. Its awful hot below decks. Most of the crew is suffering with heat rash. Mine got infected on my arm, blisters. Its not too bad and I'm having it taken care of in sick bay. Water was nice & calm today.

Thursday/Friday July 27, 1944

Made strikes on Yap. Our other task force is bombing Palau where the new **Hornet** *made its first strike in the Pacific. The* **Yorktown** *lost a plane. Went down over the Island [off] Yap. We lost a SB2C, both pilot and gunner went down.*

Friday/Saturday July 28, 1944

Saw pictures in the mess hall of the attack on Pearl Harbor. It took place Dec. 7, 1941. Sunday morning the Chaplain was saying devine [divine] services in a field near Pearl Harbor when Jap planes came. G.Q. was blasted out from the ships & the Chaplain cried "men hurry to your battle stations & God bless you." They didn't have a chance. It was slaughter. They also showed them rebuilding our fine ships that went to the bottom. Now its our turn to let the Japs take a beating & our air group has been giving them a darn good one. Bombed Yap again today.

Saturday/Sunday July 29, 1944

Had two flight patrols. A destroyer hit our ship aft on fantail while coming along side, no serious damage done. We are supposed to anchor at Saipan soon to pick up ammunition. Pay slip was up today. I'm drawing two hundred and ten dollars. Something is wrong because I only expected 45 dollars. Figure they must have been paying me less in the past months & this must be back pay. Heat is very bad below decks & most of crew is suffering from heat rash. I'm taking care of my infected heat rash so it won't spread, I just have it all over my left arm. Running short on food.

Sunday/Monday July 30, 1944

Had two flight patrols. Sighted Saipan, but didn't anchor because ships were being refueled at sea most of the day. We had a movie tonight **Ten Gentlemen from Westpoint.** *Well up to this present day our air squadron has 236 Jap planes to the* **Hornet's** *credit. We have the record so far also for shooting 67 Jap planes in a day which is a top record for both shore base and sea. We got paid today. I'm going to save three hundred dollars & send it home to the folks. Still can see Saipan. Didn't anchor today. Water is rough. Just cruising around* *near Islands.*

Monday/Tuesday July 31, 1944

We refueled at sea. My three months of Mess Cook will be up soon, but I'm going to sign for another three months. No use going back to push planes because its not what I want.

'44 August

Tuesday/Wednesday August 1, 1944

Anchored at Saipan 1750. Water was rough today. Tonight I could see tracer bullets flying through the sky on Saipan & Tinian. Our new Captain came aboard. His name is Doyle. In away I hate to see our old captain go. He was a swell guy. Saipan is ours but they are still fighting on Tinian, and it won't be long before that will belong to us to[o]. We are suppose to take bombs from the Island, but the water is to[o] rough. The chow is still lousy. We had soda crackers instead of bread because they are short on flour. The crackers were stale, & once in awhile you get a couple with bugs in them.

Wednesday/Thursday August 2, 1944

Moved position of ship. We went through the gap which separated Saipan & Tinian. I could see American planes on the Island Saipan. We could also make out sounds of gun fire in the distance. It must have come from Tinian. Saipan has a huge mountain. We couldn't see the town because it is located on the other side of the mountain. Tinian more level land then Saipan. Tonight again we could see tracer bullets on Tinian. Just a few miles away from us men are dying like rats. We had to leave Saipan late at night because water was to[o] rough and no bombs could be taken aboard.

Thursday/Friday August 3, 1944

We are heading for the Bonin Islands again, but because of the rough water at Saipan we didn't take a good heavy load of bombs. Had two patrol flights today. One of our TBFs went into the drink this morning, but the pilot an[d] the gunner were saved. Had a mail call. Got our mail from a destroyer which got our mail from the post office on Saipan. Received three letters from the folks & it was good to hear from them. Today I signed up for another three months of Mess Cook. Heading full speed for the Bonin Islands. Food is still lousy, but after we finish this strike our next stop will be for supplies.

Friday/Saturday August 4, 1944

Our air group made the raid on the Bonins today. They gave them plenty of hell. Jap ships & alot of them. All Jap ships in the harbor were completely destroyed. We had G.Q. 1410. I found out the reason our cruisers left the task besides battlewagons to hit a Jap convoy near Bonins, and boy did they hit them. The **Hornet's** *planes sunk a Jap destroyer, 5 cargo ships & quite a few LST. Don't know altogether how many ships our planes sunk yet. Our TBF made a direct hit on the Jap cruiser which they sunk. The Island Chichi Jima was shelled tonight. For the day we lost one fighter, but the Japs lost many ships.*

Saturday/Sunday August 5, 1944

Cloudy & raining. We were in G.Q. condition one easy all day because our cruisers were still letting the Japs have it by shelling Island. I could see our destroyers blast four LST out of the water. The Jap LST were firing back with their 20mm guns. Today our guns opened fired for the first time. A Betty went by, had to stop firing because high in the clouds one of our task force fighters was sighted. He shot the Jap Betty down. Beano let a Betty have it while on a search party 210 miles from Japan. His pilot opened fire to[o] & then our two fighter escorts finished the job. A Harro was picked up today, but his pilot drowned. They hit the water while attacking the Jap convoy.

Sunday/Monday August 6, 1944

Left Bonin Islands with just a few bombs left. We destroyed 5 cargo ships, 4 LST, & three destroyer escorts. The Island was shelled by battlewagon & cruisers & destroyers. A gunner told me about one of the destroyers they sunk yesterday. His plane went in low when she was sinking. First Kowal from Conn. His pilot made a direct hit & blew one of the Jap gun turrets off the deck high into the air & it landed on the bridge twisted to hell. Also when his plane went in low they could see the Japs abandon ship. We are heading for Eniwetok, and will stay there for quite some time taking supplies aboard.

Monday/Tuesday August 7, 1944

Our guns did a lot of practicing firing at sleeves. Already we are getting our ship ready for Captain's inspection. Worked all day in the mess hall cleaning bulkheads, overhead, & the decks. Water has been rationed again. The chow is very lousy—lousy meat rotten & rest of food dehydraided [dehydrated]. Boy what I['d] do for some of mother's cooking right now. When we reach Aniwetok [sic] we are going to have alot of work parties. Its still very hot below decks so I've been sleeping in a hammock under an air vent since I've started mess cook. Most of crew is suffering from heat rash. Its bad when its gets infected. Today I had my first fight. Hit a wise guy twice, and he had two stitches put in his eye.

Tuesday/Wednesday August 8, 1944

New Captain took over command. Captain's name Doyle. Our captain Sample will leave us soon as we reach Eniwetok. Sample made Admiral, and I think he deserves it. He was good to the crew & thought alot about the ship & men. They awarded the pilots all the air metal [medal]. Each and every one had Jap planes to his credit. Farrel, a gunner from Hartford, Conn. got the air metal for shooting down a Jap plane. Also quite a few pilots received the Purple Heart. There are still plenty of metals to be given out to the TBF & SB2C pilots. One gunner from a SB2C I know got the Purple Heart today.

Wednesday/Thursday August 9, 1944

Early in the morning I could make out the Island Eniwetok. We passed quite a few ships anchored near shore, most of them were troop ships. Anchored at 1145. Right off the bat working parties taking loads of supplies aboard. Still working in the mess hall. The heat below decks is very bad. Wish the Air Dept. Rates will open soon. Held field day. We are suppose to have a Captain's inspection soon. Don't know how long we will stay at Eniwetok. Took on a darn good load of bombs and I'll bet alot more will be stored.

Thursday/Friday August 10, 1944

This morning I went up to the hanger deck to take a look around. I saw we are anchored next to the battleship **Penn.** *Destroyers & tankers were anchored so close together that you could walk from one ship to the other. Started liberty parties to go on the beach of Eniwetok. I'm hoping I'll get to go on the Island this time. Missed out on the last ones they had. Only reason the sailors want to get on the beach is for beer. For dinner we finally had butter & spuds served. Its been a long time since we had real spuds & butter.*

Friday/Saturday August 11, 1944

Had swim call, swimming over the side of ship. I've been busy lately keeping up with my writing & washing clothes. The chow is still lousy. Can't get use to this Navy chow. Boy do I miss my mom's home cooking. The meat we get is lousy, awful greesey [greasy] & dry with no taste to it. Still no bread except soda crackers. Most of the time you find bugs in them. Recreation on the hanger deck—basketball, volley ball & etc. Liberty call again. A couple of mess cooks went on the beach, but I still didn't get a break. Had plenty of mail calls.

Saturday/Sunday August 12, 1944

Lousy chow. Started to have real spuds & butter but still no bread & that meat we have been getting lately stinks. Made quite a few transfers off the ship mostly rated men. Captain Sample made Admiral and he is leaving us. Our new Captain is already aboard. His name is Mr. Doyle. Liberty call again, but still no luck. Read most of the papers I've got from home. One paper I got was very wet. It happen[ed] to be one which told about our task force hitting the Jap fleet. That will be something I'll never forget. Our planes had a tough time finding the ship that night.

Sunday/Monday August 13, 1944

Had a high mass this morning & sailors from destroyers & ships with no chaplains came aboard for mass. It was a swell mass & many sailors from other ships came. A surprise—good chow for a change. Meat which cut easy, pie & bread. This time we are taking on an awful big load of food supplies. On the hanger deck all I could see was bombs, torpeadoes, ammunition, & food supplies. Last few nights we have been having good movies. We have our screen on the hanger deck, but other ships have theirs out in the open. 42 of our crew transferred already.

Monday/Tuesday August 14, 1944

I'm having a tough time in the mess hall. Its not the work, but the heat & no sunlight. I think I'm going to get off of mess cook at the end of this month. Liberty call, but still no luck. Hope I'll get to go on the Island before we go out on our next raid. Still have working parties. Awful good load of bombs came aboard. This time many ships are anchored here. They keep coming in right along. When we leave Aniwetok a very tough task is waiting for us. First Palau will be invaded & then the worst part, we are going to hit the Philippines.

Tuesday/Wednesday August 15, 1944

Great surprise tonight. U.S.O. show with five girls from

Hollywood to entertain us. Its been over five months since we have seen a girl. Weren't so good when it comes to entertaining, but they were nice looking & that was plenty good enough for the crew. One could jitterbug so the band played the jive & two sailors gave her a work out. After they finished with their performance we had a good movie, **Going My Way** *with Bing Crosby. Heat in the mess hall is still terrific. Went up to the flight deck this noon time to get some fresh air.*

Wednesday/Thursday August 16, 1944

Still having work parties breaking [bringing?] alot of bombs aboard. Had two mail calls. Received a lot of letters in the last couple of days. Anthony's letter got here & also a good one from [Aunt] Lena. Holding field day. We are suppose to have an inspection of ship before leaving to give the Japs another swift kick in the pants. No liberty party because of no transportation. Still working in the mess hall. To[o] many guys hanging bars [?] and I can't get a good break. Band played before we had the movie. When the ships are lighted up it looks like a big city all lighted up.

Thursday/Friday August 17, 1944

Still work parties. Sure are taking a heavy load of bombs, this time we are getting ready for something big. Liberty party. Still no luck. Some of the boys who have already gone told me its lousy, but what makes it heaven is the 4 cans of beer ration they got. Early chow for liberty party. Held field day. Instead of movie had a happy hour. Band played & a few of the sailors entertained us. Half way through the show Executive Officer told us to hurry an[d] go to our G.Q. stations. Found out it was bogies picked up on radar on Island. Turned out to be friendly.

Friday/Saturday August 18, 1944

*Bomb work party again. The mess hall was loaded with bombs. Had liberty party again but still I didn't get a chance to go on the beach. Found out our **Hornet** got credit for sinking Jap CV carrier when task force hit the Jap fleet. Our bomber squadron got seven direct hits on her. Pilot who was picked up said he saw the Jap carrier blow sky high & sunk to the bottom. Awful hot below decks. I'm sleeping under an air vent in the mess hall. So far I haven't fell out of my hammock. The band played and we had a movie.*

Saturday/Sunday August 19, 1944

What a record. Our Island an[d] carrier is filled with Jap flags an [for] strikes on so many Jap Islands we made. Squadron has 236 Jap planes to its credit—that is, shot out of the sky & not counting the hundreds they destroyed on the ground. 9 cargoe [sic] ships, one mine layer and a Jap carrier CV. Our air group has the highest record in the fleet. In one day our fighter squadron shot down 67 Jap planes. Bombs still coming on. Took 82 1000 pounds. Also many 100 pound bombs. Band played an[d] had a movie, **Show Business** *star[r]ing Eddie Cantor.*

Sunday/Monday August 20, 1944

Early this morning G.Q. was sounded, didn't last long. Bogies picked up turned out to be friendly. All the ships blacked out, in a few seconds we were ready for action and getting on the way. Bombs still coming on. This time we are taking a very heavy load of supplies. We'll be going out to sea tomorrow only for a couple of days to practice firing runs & pick up new planes, fighters and new SB2C with four bladed props & more speed. Our old SB2C were 3 bladed & carried a 1000 pound load. The new SB2Cs will take a 2000 pound load. Three star came aboard. Getting ready for something big.

Monday/Tuesday August 21, 1944

Left Aniwetok 0645. Had regular G.Q. Was up on the hanger deck during G.Q. watching our carrier pass small Islands & many ships anchored. Flight Quarters sounded 1210. Planes did a lot of practicing firing runs and water bombs for practice hitting skee sleads. 5" 40 & 20mm firing runs. Water was rough. The **Franklin** *is along side of us. I watched her guns blast away. You could see a big fire flash from her 5" guns when they were fired. Held field day this morning. Have an escort of four destroyers. Had night firing runs with tracer bullets.*

Tuesday/Wednesday August 22, 1944

Reville [reveille] 0300. Flight Quarters 0400. Planes did alot of practicing most of the day, firing runs & dropping water bombs. Went back to anchor. Reached Aniwetok 1615 & anchored. Our planes landed on the beach, not all of them but a good many did. All of the Bakers & a few fighters. Had a mail call, but I didn't receive any mail from home. Hanger & flight deck was scrubbed. More bombs came aboard. Got the **Hornet Tales.** *It had pictures of the fighter squadron & pilots & gunners who were awarded purple heart, air metal, & the distinguish flying cross.*

Wednesday/Thursday August 23, 1944

Still anchored at Eniwetok. Next time we leave it will be for hunting Japs again. Plenty of ships & carriers are anchored here. Bombs still coming aboard. Liberty call again, but I didn't get a chance to go. Heard its not so hot. Just alot of sand. Its only a small Island with a very few palm trees. Heard it is swell for swimming. Mail call, but still I didn't receive any mail. After supper I went to see the movie. The name of the picture was **Pin Up Girl**. *It was O.K. After the movie I went up to the flight deck to get a little fresh air before going below.*

Thursday/Friday August 24, 1944

After breakfast we had mail call & I received two letters from the folks. Work party taking ammunition & 5" shells. Got one of the new SB2Cs on. Same build as the old ones except for the four bladed prop. Hope these Bakers are better then the last one. In the aerial gunner bomber squadron out of ninety men up to this date they lost 36 men. Saw Kowal on the flight deck this noon time & we got to talking about Conn. Kowal told me our air group will leave before we finish with our operations on this mission. New air group will be 11.

Friday/Saturday August 25, 1944

Bomb working party. Liberty party went on the beach again. Took our 26 new SB2Cs with four bladed propellers aboard. They were very dirty from lying on the beach. Some of my pals were on the transfer list & left the ship to go back to the States & get on a new construction. This morning three star admiral came on ship. His name was McLain. He came from the **Wasp.** *She will be with our group in task force & she will be the flag ship instead* *of us. Found out this is going to be a very big event. Also going to strike Island in Philippines before McArthur's troops invade.*

Saturday/Sunday August 26, 1944

This morning a four star admiral came aboard. It was Admiral Mitscher. He came to present our air group members medals such as the distinguished flying cross, air medal, purple heart, & the gold star to quite a few of our air group members. Admiral Mitscher said the **Hornet** *had the best record in the fleet & he said she proved to be even better than the last* **Hornet***. Took bombs on again & more supplies. This time we are getting enough to stay out for four months, but I heard some dope that this event will take at least three months, & boy that isn't hay.*

Sunday/Monday August 27, 1944

West to mass this morning. Work parties still taking supplies food & ammunition. We have quite a few CV carriers anchored here at Aniwetok: **Franklin, Independence, Wasp, Enterprise, Lexington, Ticonderoga, Intrepid,** *and* **Essex.** *Don't know the rest of the carriers but there are about 10 first line carriers anchored here at Eniwetok. Was up on the flight deck. Beano was cleaning his guns on his new plane. I helped him out a little. It was beautiful* *today. Sun was out very strong & the water was very calm. It looked like glass, the water was so still.*

Monday/Tuesday August 28, 1944

Still taking supplies. We are leaving Aniwetok tomorrow, & heading for a pack of trouble, but I'll bet the Japs are going to holy hell & will be in for an awful surprise. Went up to the flight deck to get a sun bath. Saw the plane Captains are having a tough time trying to get the planes from the beach cleaned. Boy they are really dirty. Heard that we will pick up quite a few battle wagons on our way out to Palau. Counting all the converted carriers & plus our first line carriers I say we will have close to 1500 planes which will give the Japs plenty of hell.

Tuesday/Wednesday August 29, 1944

Left Aniwetok 0545 and this time we are going to give the Japs a real swift kick in the pants. First Palau will be invaded & then we are going to hit the Philippines. Had flight quarters. Planes did alot of practicing dive bombing and fire runs. The new SB2C with four bladed props work fine. They take off like a fighter and have alot more speed then [than] our old SB2C. On the hanger decks we had loose bombs secured to the deck 2000 pounds, 1000, 500, & 100. The Philippines they figure will be a tough battle, but we are ready for them.

Wednesday/Thursday August 30, 1944

Flight Quarters, planes did alot of practicing & firing runs. Ships also had firing runs. Lousy weather. Raining and cloudy. Tomorrow will be my last day on Mess Cook. Air Department Rates are still held up. Wish they would open so I can try for 3rd class. Served late chow. Our night fighters went up for patrol duty. Got a crews letter from the Captain telling about the ship's anniversary. The **Hornet** *is one year old today. This letter was passed by the Censor board so I sent it home to the folks. Hope I'll get into a shop now that I'm getting off Mess Cook.*

Thursday/Friday August 31, 1944

Flight Quarters, our planes did a lot of practicing. Had to work in mess hall to show my relief what to do. Tomorrow I get out of the mess hall for good. Its four months today I've been on mess cook. Its not tough work, except for the heat and staying below decks. Water was nice and smooth. It looked just like glass. Night fighters went up on a patrol search. It won't be long before we will reach Palau. Destroyer came along side for fuel. Still don't know if I'll get in a shop or not. Short on plane pushers. I'm keeping my fingers crossed.

'44 September

Friday/Saturday September 1, 1944

Started the new month by getting off of mess cook. Back on the hanger deck. Still don't know for sure if I'm going back to a plane handling crew. Very short on men. The **Wasp** *is the flag ship in our task force. Flight Quarters, our planes went up for practicing. Night fighters went up during the evening for patroling. Task force is very big. Found out they changed number of task force from 58 to 38. Our air group will be leaving us. They are suppose to get off at one of the Admiralty Islands.*

Saturday/Sunday September 2, 1944

Found out I'm supposed to get in the tire shop. At the present time I'm working with the handling crews on the hanger deck. It feels good to get out of the mess hall. Refueled our carrier. Tanker came along side. Water was nice and calm. Could see the rest of task force on the horizon. Night fighter patrol was canceled. Saw Beano on the hanger deck, and he gave me some inside dope to what we are heading for, and it looks like its going to be a tough fight. First Palau will be invade[d] September 15. Worst of all we are going to hit the Philippines.

Sunday/Monday September 3, 1944

G.Q. this morning. Had to sweep down the hanger deck. Alot of crew is still suffering from the heat rash. After working in the mess hall so long I'm use to the heat and its doesn't effect me now. Found out Mike had a fight with wise guy Smokey. Mike hit him so hard he broke Smokey's jaw, and also cut his eye. Mike had to have his hand taken care of because he swung so hard he fractured his hand. Smokey has to stay in sick bay for six weeks. Flight quarters 1245. Was up on the flight deck when planes took off. Saw our fighters test new chemical tank bomb.

Monday/Tuesday September 4, 1944

Finally got a break. My division officer assigned me to the tire shop. Start work there tomorrow. Was lucky to get such a break because they are short of plane pushers. Had flight quarters, patrol flight. Our squadron is suppose to leave for the States by the end of this month. Frenchy promised me he would see the folks for me, and tell them I'm O.K. Sure wish I was going back home. Its been a long time since I was home last. In a couple of days we will reach our destination, and strike Palau. Out here our thousand of troops waiting.

Tuesday/Wednesday September 5, 1944

No more plane pushing and mess cook. Started to work in the tire shop. Its a swell shop, and my pal Reed (who was on mess cook) and I work together fixing tires. We changed a few fighter tires. Had a patrol flight. We were suppose to have a fighter sweep on Palau, but they canceled it. On Aug. 7 [meant Sept. 7] we make our first strike on Palau. If the rates open up I'll have a good chance to go up for Aviator Machinist Mate 3rd. Altogether there is six men in the shop. Reed & I fix tires. The other four guys are in check crews. They check planes before take off, and after landing.

Wednesday/Thursday September 6, 1944

Regular G.Q. My G.Q. station is in the tire shop. **Wasp** *fighters made a fighter sweep on Palau. Little enemy resistance was accounted for. Tomorrow our planes will bomb Palau. Flight Quarters, patrol flights. Didn't do much work in the shop so I washed some of my clothes. Went on watch 2400 to 0400. Palau will be invaded Sept. 15. Part of task force which includes the* **Hornet** *will be striking the Philippines soon. Like my new job in the tire shop. Altogether there is six of us and one chief. I'll have a good chance to go up for 3rd now.*

Thursday/Friday September 7, 1944

Regular G.Q. this morning. Flight Quarters sounded 0400, our planes made four strikes on Palau. Planes met light resistance. One of our fighters was shot down over Island. No crack ups during operations. Enemy resistance just accounted for a very few spots of Palau. No enemy planes or ships were sighted. Reed and I changed a fighter tire this morning. Planes had no flats while in operations so we didn't have much work to do. Weather cloudy, and little rain. Place our planes hit at Palau was Anguai. Studied my course book. Figure the rates will be open again soon.

Friday/Saturday September 8, 1944

Regular G.Q. this morning. Flight Quarters 0450, seven flights were made patroling for enemy subs, planes, or ships. Lost one fighter operational. Don't know if pilot was saved. Reed and I fixed five flats. Our captain made a speech. He said tomorrow we will hit the Philippines. Ship expects to see plenty of action. The Japs are supposed to have many airfields on the Philippines. We started to getting ready for strike. Orda [ordinance] department started to work midnight loading bombs, and supplying planes with rounds of ammunition.

Saturday/Sunday September 9, 1944

Made strikes on the Philippines. Our first flight took off early in the morning. Bombs and ammunition was ready and waiting to be put into good use. All of us were waiting for the results for the first strike. We were ready and expected the Japs to come out and fight, but this is all that happened: Converted carrier **Bellawoods**

[Belleau Wood] shot down two Betties, a Jap convoy of 52 Japanese vessels was destroyed by our task force cruisers, 32 loaded cargoe ships and twenty were sampans. All were sunk. Reed and I changed 12 flats. Our planes made 7 strikes on Mindanao.

Sunday/Monday September 10, 1944

Flight quarters 0410. Still a surprise no Jap resistance at Mindanao. Only had three flights and then Flight Quarters was secured. Our ships went in as close as 15 miles from the shores of Mindanao. We could see the Island. It was a very big Island. Reed and I fixed one fighter tire. Frenchy came to see me. The squadron will be leaving soon, and he is going to see the folks and let them know how I am. Lousy day. Cloudy and rain. Had mess 1600 on forecastle. Our strikes on the Philippines have turned out to be a very big surprise. Very little enemy resistance is accounted for.

Monday/Tuesday September 11, 1944

Refueled this morning, tanker came along side 0810. Water was very rough. Didn't make any strikes. Found out we are going to hit Luzon, other huge part of the Philippines. It is believed that the Jap Fleet may be anchored at Manila. Also expect to see Jap airpower resistance. If this strike turns out to be like Mindanao, well then the Japs better call it quits. Flight Quarters 1510, sent up one patrol squadron. No work in the ship, so I studied my mech book, and also washed some of my clothes. So far the Japs have been getting a swift kick in the pants.

Tuesday/Wednesday September 12, 1944

Planes took off 1st hop 0730, target was Negros. Pilots reported by radio, "alot of rats out here," meant Jap aircraft. For the day, our planes destroyed over fifty planes on the ground, and also shot down 15 out of the sky—1st flight 8, 2nd 7 which made a total of 15 Jap planes shot down. Made a strike on Sebu [Cebu], and our planes sunk two freighters and one sampan. Lousy day, cloudy. Had a mail call, picked up mail from destroyer. A couple of our planes were banged up from Jap AA on Island. TBF hit, one gunner wounded, but he is going to be O.K. 9000 [4000?] landing on **Hornet** *was made today. Pilot made force landing on Island of Philippine.*

Wednesday/Thursday September 13, 1944

Planes took off early. Strikes for the day were: A-Bacolod Field, B-Duncognete Field, C-Negros Field, D-Cebu [illegible] Airfields. **Hornet's** *record for today was five Zekes, and 3 Nates. About 12 were destroyed on* *the ground. One of our fighters came in for emergency landing. His landing gear wouldn't release. Came in with landing gear folded, and he made a beautiful landing. News broadcast. Admiral Mitscher announced our damage up to Sept 12: 72 Jap planes shot out of sky, 118 damage[d] on the ground. Many freighters and sampans were sunk, I forgot number. Our pilot told us about this force landing yesterday.*

Thursday/Friday September 14, 1944

Early Flight Quarters, on the 1st strike our planes sunk a Jap destroyer. They made three 1000 pound direct hits, and also four five hundred pound direct hits. The Jap destroyer sunk in less then four minutes. Pilot of bomber squadron told us over speaker how they sunk this Jap destroyer. Said when they were ready for the kill, **Wasp** *planes asked "What's the position of this Dog Dog?" Our bomber squadron leader radioed back "If you wait about five minutes, I'll tell you where there [their] position was." Then they went in for the kill, and the Jap destroyer sunk in four minutes. Con[tinued] Sept 16*

Friday/Saturday September 15, 1944

Invasion of Palau. At dawn, part of our task force is standing by. **Hornet** *assigned patrol, only stand by to give air support if needed at Morolai [Morotai?]*

(Halmahera). Stand by to strike aircraft and shiping at Menado [sic]. Two of our fighters while on Patrol shot down a Jap Betty. Reports on invasions. Palau, our troops are meeting stiff enemy resistance. McArthur at Morolai, no resistance at all. Received a copy of **Hornets Tales.** *Figure it won't be long before the Philippines will be invaded.* **Hornet's** *paper shows score on Island.*

Saturday/Sunday September 16, 1944

Continue[d] from Sept 14. Left Negros and Sebu & hit Mindanao again. Operation for the day: Sweep Dav[a] o Gulf, Strike A Dav[a]o Area, Strike B [illegible], Strike C [illegible]. Score: 1 Jap destroyer. Our losses: 1 F6F operational, Sept 16, 1944. Still stand by at Morolai. Flight Quarters, had a few patrols. Reed and I worked with deck crew on SB2C. Destroyer came along side and refueled. Night fighters went up. When coming in for a landing, one of our night fighters crashed on the flight deck. TBF crack up was pushed over side on number two elevator. Tomorrow we refuel our carrier.

Sunday/Monday September 17, 1944

Early this morning tanker came along side and we refueled. Could see most of ships in task force refueling. Water was very calm, and it was a beautiful day. Painters added to our score on [ship's] Island: 5 more Jap cargoe ships, 26 more Jap planes and one Jap destroyer. Altogether now record is 257 Jap planes shot out of the sky, 14 cargoe ships sunk, one CV Jap carrier sunk, 1 Jap destroyer sunk, and 1 Jap mine layer sunk. Also many Islands names and number of strikes. The **Hornet** *has highest record in the United States Fleet, both shore base and sea record.*

Monday/Tuesday September 18, 1944

Regular G.Q. this morning. Water was very calm. No strikes. Had Flight Quarters late noon, planes took off on a few patrols. Reed and I fixed one flat. Not much work to do in the shop, so I washed clothes, and wrote some letters. Our next strikes will be Manilla [sic] in the Luzon. After this mission our air group #2 will be leaving the **Hornet** *and air group #11 will come aboard. Destroyer came along side. Took aboard six new replacements, 3 pilots & 3 gunners. They are assigned to VT squadron. Also took plane replacements aboard.*

Tuesday/Wednesday September 19, 1944

Regular G.Q. this morning. Held field day in the shop. No work to do so I wrote a few letters and washed some clothes. Tonight for supper we had a surprise—a can of beer with our chow. The beer was nice and cold. Flight Quarters, had a few patrols. Tanker came along side and we refueled. A yeoman friend of mine told me the Air Dept. Rates might open again soon. I sure hope they do so I can take my final exam. This noon I went up to the flight [deck] to get a little sun burn.

Wednesday/Thursday September 20, 1944

Regular G.Q. this morning. Reed and I fixed one fighter tire this morning. This noon I wrote a few letters. Our Captain made a speech tonight. Told us tomorrow the **Hornet's** *planes will strike Manila Bay in the Luzon. He said Admiral Mitscher is giving the* **Hornet** *&* **Wasp** *the privilege to hit the Japanese shipping at Manila Bay. Rest of task force planes will hit most important airfields on the Luzon. After finishing strikes on the Luzon our air group is getting releaved [sic], and our new Air Group will be Air Group #11.*

Thursday/Friday September 21, 1944

Early this morning our planes took off to hit Manila bay, and did the **Hornet's** *planes hit them. Here is our score for the day. Strike 1A: 9 Tonys, 1 Zeke, 3 large AKs sunk, 1 large AO sunk, 3 med. AKs sunk, 2 small ships sunk, 5 med. AKs damaged, 7 small ships damaged. Our losses—none. Strike 1B: 1 large AO sunk, 2 large AKs sunk, 3 med. AKs sunk, 2 med. AKs damaged. Our losses—none. Strike 1C: 1 large DD sunk, 1 large AO, 1 large AK, 3* med. AK damaged. Our losses—none. Strike 1D: 1 small AO sunk, 1 large AP, 7 med. AKs damaged, C.A.P, 1 Judy. Our loss—none. For the day 11 planes out of sky shot down, 17 ships sunk, 27 damaged, on Strike 1D we lost a TBF & crew. Score changed, definitely sunk eight cargoe ships instead of 16.

Friday/Saturday September 22, 1944

Early this morning our planes took off to hit Manila Bay for shiping. Made only two strikes because of bad weather. For the day our planes sunk two ships, damaged five, and while over target shot down four Tonys. This morning our night fighter shot down 1 Juddy which makes total for the day 2 ships sunk, 5 Jap planes shot down. Today our task force was attacked by five or six Jap planes. The **Hornet's** *guns were firing like mad at a Jap dive bomber. This Jap bomber droped quite a few bombs but missed. While in a dive, he strafed our flight deck and killed one of our gunners, injured my pal Sherman very badly, and wounded three more members of our crew. One of our cruisers flashed two, a destroyer flashed one. Don't know if the rest got away.*

Saturday/Sunday September 23, 1944

This is one day that was hell, real hell. Eddie Sherman my pal died early this morning. They had funeral services for the gunner Speni who died yesterday. I went to Sherman's funeral 1630. It made me sick inside to see Eddie's body slide of[f] the board and hit the water. Hope God will look after him and give him ever lasting peace and rest. Refueled today. Tanker came along side. On the other side of tanker the cruiser **Boston** *refueled. Mail was takened off of tanker, and we finally had a mail call. Received pictures and five letters from the folks. It sure was good to hear from them. Tomorrow squadron #2 makes their [their] last strike off the* **Hornet.**

Sunday/Monday September 24, 1944

Our squadron #2 made there [their] last strike off the **Hornet** *today. Our fighter squadron lost one fighter & pilot over Jap airfield Negros Is. The Islands our planes attacked today were Coron Island, Negros, and Panay Is. Results on strike: our planes damaged a Jap cruiser, destroyer, and badly damaged two cargoe ships. The* **Wasp** *planes sunk two Jap destroyers. Our fighters helped the* **Wasp** *planes strafe, but the* **Wasp** *got the bomb hits and gets credit for sink[ing] two Jap destroyers. Had mail call and I received seven letters from home. Captain made a speech. Told us our ship did very well on strikes over Philippines.*

Monday/Tuesday September 25, 1944

Heading for Manus Island. Our Air Group #2 is getting ready to leave the ship at Manus in the Admiralty Islands. Our new Air Group will be Air Group #11. Flight Quarters, one patrol. Marines were practicing firing runs with machine guns on the flight deck. The painters are working on the [ship's] Island to complete score of ship. Without counting Jap island bombed, just Jap planes shot out of sky and ships sunk here is Air Group #2's record on the **Hornet:** *272 Jap planes shot out of the sky, one CV carrier sunk, 1 mine layer sunk, two Jap destroyers sunk, and two cargoe ships sunk.*

Tuesday/Wednesday September 26, 1944

Had two patrols. Air Group #2 made their last landing on the **Hornet.** *Captain made a speech. He said Admiral Mitscher from the* **Wasp** *watched the last plane of Air Group #2 to land on our carrier. He said they did their job very well, and I was proud to have them with our task group. We will miss Air Group #2. Yes we will miss them. We had some swell pilots & gunners. I hate to see my pals Beano, Frenchy, Don, Kowal, and the rest of the gang leave, but in another way I'm happy to see them get a break which they more than well deserve. The flag staff will leave also at Manus.*

Wednesday/Thursday September 27, 1944

Quarters on the flight deck to see pilots receive the Air Medal, Gold Star, or Distinguish Flying Cross. Seven of the crew which were wounded from strafing while we were attacked by a few Jap bombers received the purple heart. No flight quarters. Had a movie and the band played. Pilots most of day were taking pictures of score on the Island. On the flight deck some of the crew were playing volley ball. Held field day in the shop, didn't have any tires to fix so I wrote a few letters. I saw Kowal and he told me he is going to see the folks for me when he is back in Conn. on his 30 day leave.

Thursday/Friday September 28, 1944

Reached Manus Island 1235 and anchored. Manus was a very big and beautiful Island. It was raining very hard today. As soon as we anchored, work parties began loading bombs, ammunition, and supplies. When mail was taken aboard, at least fifty sacks, it was pouring cats and dogs. Most of my mail was spoiled from the rain. We had two very long mail calls. Some of it was dry, but the home town papers I got were soaking and wet. Eddie Peabuddy [sic] and a few other musicians entertained us tonight and it was a swell show. Also had a movie.

Friday/Saturday September 29, 1944

*Our Air Group #2 left us this morning. I saw Beano, Kowal, & Frenchy before they left the **Hornet** for the States. We all hate to see our swell air group leave us. Executive Officer Durfeldt made Captain. He also left the **Hornet** today. Liberty party went to the beach on Manus Island, I may get a chance to go ashore tomorrow. I sure hope so. Its been a long time since I've had my feet on solid ground. Captain made a speech to liberty party. He said part of the island was suffering from Malaria, and for them to be careful an[d] to keep their sleeves roll[ed] down. Rain again.*

Saturday/Sunday September 30, 1944

Had a liberty party. My section 2 rated liberty, and I went on to the beach. Before today my last liberty was in Pearl Harbor March 5. It sure was good to get my feet on solid ground. We went to a recreation camp on the Island. Each man rated four cans of beer at the canteen. All you could buy was peanuts & smokes. The beach for swimming was fine so I went in for a swim a few hours. Boy what a rain storm we had on the island. Since we've been anchored at Manus, its been rain, rain, and more rain. An L.C.I. took us ashore and we left the ship 1015. On the way back we left the Island 1430. We had a movie tonight. Brought alot of bombs aboard.

'44 October

Sunday/Monday October 1, 1944

Went to mass this morning. Liberty party section 3 went to the beach. Was on a working party essembling [assembling] propellers. Still taking bombs and plenty of ammunition aboard. Our new squadron #11 is about all squared away. Rain again, but it wasn't much of a storm & it lasted only about one hour but all day it was cloudy. The mainland of Manus Island they say is very mud[d]y. Our chief was on the main land & he told all about it. Most of it is jungles, and he said the men stationed there have to wear high hip boots because its so mudy there. Had a movie. We leave Manus tomorrow.

Monday/Tuesday October 2, 1944

Left Manus Island 0745. Out to sea with new Air Group #11. This morning I was on a work party hauling oxygen tanks up to the oxygen shack, and also had to help secure wing on overhead of hanger deck. Washed some clothes. Had flight quarters 1500, new air group took off our carrier for the first time. When they came back for landings it was a mess. They just kept circling around, and being flagged off. Well they finally landed them all, two flats. Reed and I are going to fix them tomorrow morning. Guns on ship had firing runs 5" 40 mm & 20 mm. The ship is loaded with bombs and ammunition.

Tuesday/Wednesday October 3, 1944

Flight Quarters 0710, squadron took off this morning. They did alot of practicing, firing runs, and droping water bombs on skee slead. Our new air group's combat in the air was good. They made some swell hits on target skee slead. When the flight came in for landing it was much better then yesterday. Some of these new pilots have to get use to landing on a carrier. Altogether, Reed and I fixed five tires which two were from last night. So far I do not know where our new Air Group 11 will make its first strike, but I know the first will not be the Philippines.

Wednesday/Thursday October 4, 1944

Flight Quarters 0700, planes took off. Did a lot of practicing droping water bombs, and had firing runs. So far Air Group 11 did very well with their practice runs. As for landings they are also improving. Reed and I had a busy day. Had to change four tires and also I had to bring the bad tires to the issue room for replacements. Water was kind of rough, and the ship bounced around a little. Had a mail call, but I didn't receive any letter. Since we have a lot of new seamen, rated men & aerial gunners, the chow line is nice and long again. Washed some clothes this morning. Secured from Flight Quarters 1810.

Thursday/Friday October 5, 1944

Flight Quarters 0730. Water was very rough and we're heading into a bad storm. I could see our task force ships from the flight deck bouncing around in the water. Its real[l]y a sight to see. Crew was ordered to keep away from the for[e]castle beca[u]se the bow of the ship is ploughing its way through the waves and water is splashing all over the forcastle. Planes only took off twice, and what a job they had coming in for a landing with the ship bouncing up & down. Today Reed & I fixed six flats. Because of this storm, when the planes came in for a landing they hit the flight deck hard.

Friday/Saturday October 6, 1944

Early Flight Quarters 0430, plane pushers took down night fighters. The storm is worse then [than] it was yesterday. Once in a while a wave splashed as high as the flight deck. Most of the hops were cancelled because of the bad weather, just had two patrols. Reed and [I] fixed four flats. Held field day in the shop, because of inspection. Lt. gave our shop a 4.0 rating. We had the place shining. Was up on the flight deck watching the ships bouncing around. Boy one jeep carrier I saw tip so far that I thought its flight deck would go under water. Had a movie tonight at number two elevator. Name of picture: **Breakfast for Two** *B.S.*

Saturday/Sunday October 7, 1944

Hit a typhoon. Ship bounced around like a top. This morning we ran into a very bad rain storm. Our task force just missed a hurricane a few hundred miles away. No Flight Quarters. All hands were warned to keep off the forcastle, and the roller curtains were closed all day. When the wind storm qui[e]ted down a bit I went out on the flight deck. I could see our ships plowing their way through the storm. One Jeep carrier I saw waves reached as high as its flight deck. Tonight the storm quited down a bit since the wind slacked. Held a field day in the shop. Don't know where we are going yet.

Sunday/Monday October 8, 1944

Refueled task force 0800. Water this morning was still very rough, but about noon time it started to calm down a bit. While refueling I watched the tanker take a beating with the waves sloshing its main deck. On starboard side of tanker a cruiser refueled, took mail off ship and transferred it to tanker. Four sacks fell in the drink. Tonight Captain made a speech. Told us our destination first will be the southern tip of Japan Island Nansei Shoto. I t[h]ought our air group would strike an easy target the first time, but I was wrong. Sure wish we make out good on this strike.

Monday/Tuesday October 9, 1944

Flight Quarters 0630, sent up three patrols. Water was kind of rough, and believe me Reed & I had quite a day. This morning when planes came in for a landing thirteen flats. Reed & I had to take all these tires apart and exchange them for new tires & tubes. Also six wheels banged to hell, and had to be exchanged. One SB2C & one F6F cracked up on the flight deck. Captain [at] 1915 made a speech. Told us tomorrow we strike, and right now we are only four hundred miles from our target, and so far the enemy doesn't know we are coming. Sub was spoted by one of our destroyers, but it turned out to be a friendly. Hit target early tomorrow.

Tuesday/Wednesday October 10, 1944

Early Flight Quarters, first flight fighter sweep at dawn. Very little enemy resistance was found, and during the day no G.Q.s. Here is the schedule. Our planes hit Naha and Miyoko. Our planes made four strikes. Damage they did was two Med. Freighters sunk, and one small Freighter sunk, damaged one small tender and 3 Med. Freighters, shot out of the sky 1 Jap bomber, damaged on the ground at least 24 planes. For complet[e] task group score: sunk 1 Sub tender, 1 destroyer, 1 cruiser, and 4 freighters, damaged 12 Med. freighters, and 17 small ships, one destroyer beached, fourteen Jap planes shot out of the sky.

We lost one pilot & fighter.

Wednesday/Thursday October 11, 1944

G.Q. sounded 2415 [0015], bogies, 3 groups. Secured from G.Q. 0110. Hit the sack. G.Q. sounded 0315, secured 0345. Our planes hit Luzon. Sweep Luzon Aparri airfield. Fighter sweep went off 1210, came back 1320. Little enemy resistance was found. Damaged all aircraft on field. Early this morning task group refueled. Tomorrow our planes strike Formosa. Its a big surprise we are run[n]ing into little enemy resistance so close to Japan. Fixed one bomber tire. Sent up two patrols. The tanker w[h]ich refueled our carrier had two Jap planes to its credit. It only takes one bomb hit to sink a tanker. 2000 [hours], one of our night fighters shot down a Betty.

Thursday/Friday October 12, 1944

Had Flight Quarters at the break of dawn. Made four strikes on Formosa. Our ship was only 150 miles away from the China Coast. Ran into a pack of trouble tonight. We had G.Q. 1910. At 2015 task group on the port side opened fired [sic]. I was out on the cat walk watching the fire works. I saw three Jap planes burst into flames one right after another in a few seconds they were burning in the water. We had fighters up 1645. Plenty of dog fights, but as yet don't know the results. Three of our fighters came in for landings 2335. Ships sent up a huge smoke screen. It was reported **Wasp** *night fighters shot down the last two Jap planes near task group.*

Friday/Saturday October 13, 1944

A day to remember. ~~We took three torpeadoe misses & two bomb misses tonight.~~ *[Crossed out by Carmen.] G.Q. was sounded 0305. Guns opened fired. Task group on port side sent up two of our night fighters. Made four strikes on Formosa. 1700 G.Q. 1810 I saw Jap Betties making a torpeadoe run for us. Our screen open fired. I saw Betties blast right out of the sky on starboard side. On our port side the* **Hornet** *opened fired and shot down two Jap Betties. One made a suicide attempt and landed about 100 feet away from us. Altogether seven Betties were shot down by our ships guns. Damage done: one of our cruisers* **Canberra** *got a direct torpeadoe hit. One of our Marine gunners got hit in the arm from S.M.*

Saturday/Sunday October 14, 1944

Captain told us this morning one of our cruisers was hit and another cruiser was towing her. She is only making five knots. Task force is standing by to protect her. The Japs know this, and they will be coming in to get us. So for us to get ready for fire works. 1600 G.Q. sounded, sent up fighters, plenty of dog fights. Three of our fighters

*was shot down. Our fighters shot down 19 Jap planes.
A cruiser was sent to fill our screen. [illegible] ships
were attacked. 0745 on the port side I saw a Jap plane
burst into flames & hit the water. The* **Hornet** *got three
torpeadoe misses and two bomb misses. Our guns blasted
most of the night. Cruiser* **Viston** *[?] got a direct got a
direct torpeadoe hit.*

Sunday/Monday October 15, 1944

*Destroyers kept coming in to refuel from our carrier,
but G.Q.s kept being sounded, and fighters kept going
up. Captain made a speech. He said 100 of our B-29[s]
blasted Formosa last night. Left the two cruisers which
were being towed with a few Jeep carriers and a
protecting screen. Our task group put on 25 more knots.
Tomorrow we will refuel. I don't know the exact score
of Jap planes our fighters shot out of sky in the so many
dog fights in the last three days. Also exact damage we
done on Formosa. We saw plenty of fire works. Today
our* **Hornet** *shot down a Judy just over us. Droped two
bombs both near misses. Our fighters shot down 2 Judys.*
Viston *[?] part of crew abandon ship.*

Monday/Tuesday October 16, 1944

*Refueled today. Hornet News Flash. Damage our task
force did on Okinawa Islands: Sunk 1 Destroyer Escort,*

4 small Subs, 14 cargoe Ships, 1 Submarine, 1 oiler, 25 small ships, 411 barges and small crafts. Probably sunk 10 small ships, 1 Minesweeper craft. Damaged 8 cargoe ships, 1 Med. Landing ship, 1 Light mine layer, 10 small ships. In for a pack of trouble. 1810 [hours] Captain made a speech. He said our subs sighted the Jap fleet heading in to finish off our two damaged cruisers & standby task group. Our Captain said our strike on the Philippines have been canceled and we are now heading straight for the Jap fleet. Boy are we in for plenty of fire works.

Tuesday/Wednesday October 17, 1944

Looks like the Jap fleet fled again. No sign of them. Wish they would come out and fight. Don't know where we are going now. Today's Hornet News Flash: Halsey's force from Oct. ninth to the twelfth, brought the grand total including Friday's to 616 Japanese aircraft damaged or destroyed and 227 nipponese ships and small craft sunk or damaged. Tokyo Rose claimed 23 American ships had been sunk or damaged. I know for a fact old Tokyo Rose is a darn good liar. We did loss [lose] quite a few planes, and two of our cruisers were hit. Also two destroyers hit, but as for the Japs sinking 23 of our war ships, that's what I call a hell of a lie. Our ship's gunners have four Jap planes to their credit: two Bettys, one Tony, one Judy.

Wednesday/Thursday October 18, 1944

Hit Luzon. Sent up two fighter sweeps. For the day our fighters shot down 19 Jap planes. Our losses were two fighters, one over the target. The other pilot was badly wounded. He droped in the drink while circling for a landing. So far the fighter squadron lost 11 pilots. G.Q. was sounded, Bogies 1845, 20 miles. G.Q. sounded 1930. Secured from G.Q. Sent up one strike. Hit airfields on Luzon. Our planes damaged 21 Jap planes on the ground. Over the radio, Tokyo Rose is still telling alot of lies how the Japs sunk our fleet and of the thousand [s?] of American sailors that died & etc.

Thursday/Friday October 19, 1944

Made four strikes on Luzon, first at dawn. Sent in a fighter sweep over target, did heavy damage on Nielson Field. Sent up two fighter patrols. Our losses for the day were one SB2 & one TBF over target. Refueled two destroyers which came along side our carrier. Took a few bags of mail off of carrier. Had one mail call, but I didn't receive any letters. Reed and I are plenty busy fixing tires. We had to fix eight tires today. Tonight over the news broadcast on ship, we heard the invasion of the Philippines starts tomorrow. MacArthur's troops will hit the beach.

Friday/Saturday October 20, 1944

Sent up a fighter sweep over East Leyte. MacArthur's troops landed this morning 1000 on the Island Leyte. The Japs were believed to be fooled. We all figure MacArthur's troops would first hit the beach on Mindanao. Our battlewagons, cruisers, & destroyers bombarded the Island Leyte to support the invasion. Reports we had of the invasion so far were very good. Our troops landed on the beach with very little enemy resistance. MacArthur landed many thousands of our boys on the beach. Hope tomorrow they will give the Japs a swift kick in the pants.

Saturday/Sunday October 21, 1944

Refueled early this morning. Lousy day rain and very cloudy. Destroyer came along side this morning and 33 bags of mail was taken on. Had only one mail call and I only received a few home town papers. Held field day in the shop. As for where we are going, so far I do not know. Had a movie on the hanger deck. Had fire call. The movie was stop[p]ed for about a half hour. Fire was put out and the movie started again. Over news broadcast tonight we heard the reports of the invasion of the Philippines. So far very little enemy resistance. Our Captain read us a letter from the President praising the work of the Navy and MacArthur's troops while invading the Philippines.

Sunday/Monday October 22, 1944

It is with great pride that CINCPAC transmits the following message received from our President. Hornet News Flashes—Personal from the president: the country has followed with pride the magnificent sweep of your fleet into enemy waters. In addition to the gallant fighting of your flyers we appreciate the endurance and super seamanship of your forces. Your fine cooperation with General MacArthur furnishes another example of team work and the effective and intelligent use of all weapons. To the officers and men of all services who have carried the fight to the enemy, and to those who have planned and have supplied the needs of the fighting forces through the years, is due the credit for the situation which prompted the Commander-in-Chief of the Army and Navy to send his message to all those officers and men, "Well done."

Monday/Tuesday October 23, 1944

*New carrier with our task group U.S.S. **Handcock** [Hancock]. Some of my pals are on the U.S.S. **Handcock**. ~~They had a 25 seamen draft off the **Hornet** about four months ago and three of my pals left with this draft. They were assigned to the U.S.S. **Handcock**. Now they are out to sea with the fleet again.)~~ [Crossed out by Carmen.] Sent up two patrols. Held a field day scraping catwalk just out side of shop. Had a mail call,* *but it was just packages, and newspapers. I got five copies of the home town paper. So far MacArthur's troops on the Philippines hasn't run into much any enemy resistance, but sooner or later its going to be a hell of a fight on the Philippines.*

Tuesday/Wednesday October 24, 1944

We were supposed to anchor at Palau tomorrow. Was all set to make a strike at Yap before entering into port at Palau. Strike was cancelled. Captain made a speech. He said Jap fleet was spoted again and they believe she is headed for the Philippines so our anchorage is cancelled. Changed our course and heading for the Philippines. Captain said to get ready to see action. I sure hope we do catch up with the Jap fleet this time. Held field day. Had a zone inspection. Movie **Action In the North Atlantic.** *Gangway was all set to lower for anchorage, but during the day it was unrig[g]ed.*

Wednesday/Thursday October 25, 1944

A big battle with the Jap fleet is taking place at the Philippines. Task For[ce] 38.2, 38.3, 38.4 is battling with the Jap forces. Last night a surface ingagement [engagement] took place. Today we 38.1 went into action with the Jap fleet. We rescued a group of CVEs being shelled at from Jap battlewagons, cruisers, & destroyers.

Sent up two strikes. This Jap task group consisted of 3 Jap battleships, 4 cruisers, and 10 destroyers. Our fighters shot down one Zeke. We drove this Jap task force away. Our air group got bomb hits on two battlewagons, and a fish hit on a Jap cruiser. Our lost eight SB2Cs missing.

Thursday/Friday October 26, 1944

The Jap fleet our planes hit yesterday was heading for San Bernadino Straits trying to escape. Our air group got another crack at them, and got a bomb hit on a cruiser & destroyer. Our Captain told us the rest of our fleet did very well with the battle with the Jap fleet. They sank so far at least 17 ships, four of which were Jap carriers. Our loss as I know so far the U.S.S. **Princeton** *& another converted carrier sunk. One destroyer got a bomb hit and also a cruiser. Our subs hit & damage[d] Jap battlewagon and got five torpeadoe hits on it. The Japs have taken a heavy loss & are doing their best to flee. Torpedo fell out of TBF while landing and skid[d]ed up the flight deck. Threw it over side.*

Friday/Saturday October 27, 1944

Refueled early this morning and heading for Ulithi. 26 of our fighter pilots are leaving us for replacements on other carriers. The first night battle with the Jap fleet 150 Jap planes were shot down by our carrier fighters. Here is how the Jap fleet appro[a]ched the Philippines: first force of three battlewagons, two cruisers, and about thirteen destroyers sighted south of Mindanao. It was attacked by carrier aircraft and torpeado hits were made on battle ships & cruisers. Second sighted in the Sulu Sea two battlewagons, two battleships, one cruiser, four destroyers. Third Southeast of Formosa. Oct 23, the battle with the Jap fleet begun.

Saturday/Sunday October 28, 1944

Heading for Ulithi. Will reach Ulithi tomorrow. On the U.S.S. **Princeton** *her captain and 133 officers, 1227 Enlisted men were saved. Found out the U.S.S.* **Franklin** *got a bomb hit Oct 26 while attacked by Jap planes from the Philippines. The* **Enterprise** *is with the* **Franklin.** *MacArthur's troops are progressing fast on the Philippines. The Japs are put[t]ing up a heavy fight at the Philippines. Once we get control of the Philippines we are well on the way to licking the Jap. The Japs claimed over the radio the* **Lexington, Wasp,** *and* **Hornet** *were sunk. Boy that's a laugh. So far we haven't got [a] scratch.*

Sunday/Monday October 29, 1944

Entered port Ulithi. Anchored 0845. The two cripple[d]

*cruisers which were part of our screen the **Ustin** [?] and* **Canberra** *are anchored here to be repaired so they can make speed and head for Pearl Harbor. While our ships pulled into the bay to anchor, surviers [survivors] from the* **Ustin** *[?] and* **Canberra** *cheered our ships which anchored close by. The Jap fleet took a beating. I saw two hospital ships, alot of our boys got hurt out there. I saw a destroyer with two of its gun mounts twisted from a bomb hit. Right away barges of ammunition and bombs pulled along side our carrier, and the flight deck was being repaired.*

Monday/Tuesday October 30, 1944

Went to the beach on liberty. The name of the Island where we went was called Pongo Pongo. We hit the beach with a landing barge. The Island was covered with cocoa nut [coconut] trees. Along the shore were native huts. On the way to the beach, we passed the **Canberra** *and* **Ustin.** *The* **Ustin** *had two torpedo hits. The carrier[s]* **Ticonderoga, Essex,** *and* **Lexington** *we passed. On the beach we had a beer party. The Island was a seaplane base. Near shore I saw a lot of PBM. Liberty only lasted six hours then we went back to the ship. An Admiral came aboard. I don't think he will stay on.*

Tuesday/Wednesday October 31, 1944

Plenty of work parties. A barge of torpedoes came along side. Taking on a very heavy load of bombs & ammunition. A few troop ships anchored out here today. Oil was taken on tonight. Liberty party went to the beach. Replacement planes were brought on. They were hoisted up by the craine [sic]. We should be leaving Ulithi soon. Had a movie tonight. Held field day. Some of the fellows in the deck division went over the side chip[p]ing paint. Held field day in the shop. I wrote a few letters, and did a little clothes washing.

'44 November

Wednesday/Thursday November 1, 1944

*Supplies are still coming aboard. This morning. I was on a work party taking food supplies below to the store room. 1345 all the food supplies were stored away, everything on the hanger deck was squared away, and now we are all set to go out to sea again. Had a Happy Hour. Some sailors from the **Ticonderoga** held a boxing match with some of our boys. The **Hornet** beat the pants off them. Some nurses from a hospital ship came aboard visiting. After the Happy Hour we had a movie. Tonight I found out tomorrow we leave Ulithi Island. Don't know our destination where we will [illegible].*

Thursday/Friday November 2, 1944

*Early this morning the **Franklin** and **Enterprise** pulled in after having an air battle at the Philippines. The **Franklin** did get a bomb hit. As the carriers pulled in our band played and we cheered them. Left Ulithi 1245. We got a machine for fixing tires. It has to be welded to the deck before we can try it. Mail call. I received two letters from home, and sis sent me Christmas cards so I can send them to my friends. A new converted carrier replacement pulled into Ulithi this morning. Captain told us tonight we are going back to the Philippines.*

Friday/Saturday November 3, 1944

*Our fleet refueled early this morning. The U.S.S. **South Dakota** is part of our escourt [sic]. Captain told us we are going to hit the southern part of the Philippines w[h]ere the Japs have alot of planes. Found out the cruiser **Burmingham [Birmingham]** was hit bad while taking on surviviers off the U.S.S. **Princeton**. Magazine room in carrier exploded and scrap damaged the U.S.S. **Burmingham** very badly. She had 250 casualties. While refueling I saw quite a few of our battle wagons are with us. This time, it looks like we are out for trouble. Had four tires to change. U.S.S. **Reano** **[Reno]** cruiser was also hit by a torpeadoe from Jap Sub.*

Saturday/Sunday November 4, 1944

*The Japs are suppose to have alot of planes at Luzon. The Captain told us we make strikes on Luzon tomorrow, also we will be in G.Q. all day. Its believe[d] the Japs have suicide pilot squadrons at the place we strike. Altogether there will be four strikes and two fighter sweeps. Our band played on the hanger deck tonight. Reed & I fixed four tires. We have two battle wagons with our screen. The U.S.S. **South Dakota** is one. I don't know the name of the other battle wagon. Our Captain believes our ships will see plenty of action tomorrow.*

Sunday/Monday November 5, 1944

G.Q. all day. The Japs have plenty of airplanes on Luzon. Our fighter squadron shot down 27 Jap planes out of sky, and a gunner in a TBF and SB2C each shot down a Jap plane which gives a total score of 29 for the day. Our losses: one F6F crashed over target, and one TBF pilot and radioman were seen bailed out over target. 38.2 [illegible] were at[t]acked by Jap planes. Three were shot down, but one Jap plane made a suicide dive at the **Lexington** *and crashed into the signal bridge. Little damage was done. Airfields were heavily bombed and well over 100 Jap planes were destroyed on the ground. Made four strikes & two fighter sweeps.*

Monday/Tuesday November 6, 1944

G.Q. most of the day. Torpedoe defense call was sounded 0309. One of our destroyers in our task group 38.1 spotted a sub. Jap sub launched three torpedoes but missed destroyer. Made a fighter sweep and two strikes on airfields in Luzon. Damaged many Jap planes on the ground. One of our SB2C came back with a huge hole through the wing about seven inches in Diam[eter]. Another SB2C made an emergency landing, no tail hook. He made a beautiful landing and didn't even hit the barrier. A fighter crashed on last hop coming in. Part of prop hit a Mech in the face. He was cut up bad, but he is going to pull through.

Tuesday/Wednesday November 7, 1944

Left Philippines early this morning. We refueled from tanker. So far I don't know what our destination will be next. Found out from our Hornet News Flash, the Japs made a raid on Saipan & Tinian Oct 13, the ~~same time~~ same day we were attacked by Jap Betties. Also part of our fleet at the Philippines was being attacked. As for the damage done it was very little. Water was very rough. Reed and I fixed a few tires and I washed some clothes. Mail was taken off the ship to tanker while refueling this morning. MacAurthur's troops still advancing in the Philippines but the Japs are putting up a hell of a fight.

Wednesday/Thursday November 8, 1944

During the day we kept get[t]ing reports of the election for president. Roosevelt won as usual. Captain made a speech. Told us we are only about 300 miles away from a tropical typhoon. Believe me, the water was very rough. No one could go out on the forcastle. The Jeep carriers with our task group was taking a hell of a beating. The **Wasp** *left and the* **Yorktown** *took her place. Reed and I fixed four tires. Still don't know our destination. Flight quarters was cancelled because of the bad weather conditions.*

Thursday/Friday November 9, 1944

Water alot calmer. Sent up two patrols. Reed & I fixed three tires. Admiral Montgomery left the **Hornet**. *He was transfered off our carrier to a destroyer, and then he was transfered to the* **Yorktown** *which will be our flag ship in task group 38.1. The flag staff will be leaving tomorrow morning to be transfered to the* **Yorktown.** *Mail was taken off a destroyer. I got three letters. Still we don't know our destination. Captain made a speech telling us Admiral Montgomery is very well pleased to have the U.S.S.* **Hornet** *in his flag group.*

Friday/Saturday November 10, 1944

A search plane spoted four Jap battlewagons, three cruisers, destroyers & troop ships heading for Leyte. They are trying to sneak troops on Leyte, but now we are wise and are on our way to hit this Jap troop convoy. Our Captain told us tomorrow morning at dawn we make the attack. All day long we have been making 27 knots. With our task group 38.1 are two battlewagons, U.S.S. **Washington** *[&] U.S.S.* **South Dakota.** *We were suppose to refuel early this morning, but it was cancelled and we are speeding back to the Philippines to hit the Japs.*

Saturday/Sunday November 11, 1944

*The Japs succeeded in landing troops on Leyte. An estimate of 35,000 Japs hit the beach. Our fleet did spot a Jap convoy consisting of four AKs, 2 PD and 2 DE. All were sunk. The **Essex** planes reached the target first. Our planes did get two torpeadoe hits—one on a Destroyer, and the other one on Destroyer Escourt. They took good pictures showing the Jap ships sinking. So far I found out we had over 300 planes replaced on the **Hornet**. Most operational losses, and up to this present date 65 pilots from the **Hornet** were killed or lost. Sent up a search hop, but no more Jap shiping spoted.*

Sunday/Monday November 12, 1944

Early this morning we refueled from the tanker. The band played on the hanger deck while we were refueling. Sent up one patrol. Didn't do much work in the shop. I could see most of our fleet refuel from tankers. 1710 [hours] SB2C flew over our flight deck, and droped a message. About an hour after this our Captain told us our orders are to head back to the Philippines to hit Manilla and Luzon Jap concentrations. We're going to make a speed run during the night and should be making strikes sometime tomorrow.

Monday/Tuesday November 13, 1944

*Early reveille 0340. Operation Schedule: 0615 to 0915 VF sweep Nichols, Neilsen, Lipo airfields. Strike 1-A Manila Bay shipping. 0900 to 1230, Strike 1-B Manila Bay shipping CAP #2 1015 to 1430 to Coord. Manila & Relieve on station 1145 to 1430 Strike 1-C Manila Shipping CAP #3 1415 to 1715 Strike 1-D Manila Shipping. 1740 CAP #4. 1 Betty shot down while going over target on Strike 1-A. Shipping 22 ships in the harbor. Fighters shot down 2 Jap planes which made a total of three planes. Fighter strafed ammunition train. It blew up. On strikes in Manila Bay our **Hornet's** planes damaged 9 ships. Over target we lost two TBFs. One was the com[mander] of the TBF squadron.*

Tuesday/Wednesday November 14, 1944

Made three strikes and hit Manila Bay & Jap airfields. Not much shipping and enemy planes found. Our air group heavily damaged two AKs and also strafed and bombed air fields. Had G.Q. 1510 to 2030. One strike was cancelled. We left the Philippines tonight, and we will refuel in a few days. After that I don't know what will be our next move. Fighter came in with no flaps. Made a good landing. His plane was shot up badly while over target. SB2C crashed on flight deck. For the day we had no losses.

Wednesday/Thursday November 15, 1944

Sent up two patrols. The **Wasp** *is still with our task group, but as for what next we are still waiting for orders. Held field day in the shop. Our Captain gave us the straight dope. A lot of tales were spreading around the ship about going back to the States, Pearl Harbor, or Sidney, Australia. Our Captain said as for going back it won't be before next spring and as for what comes next, he didn't know because we haven't our orders yet. Fixed one fighter tire. Wrote a letter home, and did a little studying for Mech 3rd.*

Thursday/Friday November 16, 1944

Left 38.1 and joined with 38.4. Since we have been out with the fleet we opporrated [operated] with the **Bunker Hill, Yorktown, Wasp** *& now we are opporrating with the U.S.S.* **Enterprise.** *The Admiral on the* **Enterprise** *is Admiral Davison. Had Flight Quarters and took on replacements. Captain told us one of his best friends is the captain of the* **Enterprise,** *and he said he wants us to continue to do good work so he can show his buddy what a fine carrier outfit we are. Also he said we are standing by near the Philippines for a few days just in case MacAurthur has trouble on the Philippines.*

Friday/Saturday November 17, 1944

Reed and I fixed five tires. Refueled from tanker, also refueled destroyers. Had mail call and I received a few more letters from home. Still standing by. Band played on the hanger deck. We were going to have a movie on the hanger deck at number two elevator, but it was cancelled because the night fighters were skedule[d] to go up. Held field day in the shop because of zone inspection. Set up the machine to fix tires. Sure hope it will work [illegible]. Good years [tire brand] are plenty tough to put on a wheel. Officer gave us 4.0 for the shop's inspection.

Saturday/Sunday November 18, 1944

Refueled a few destroyers. Gostin heard from Jessie and he showed me his letter. Hope Jessie will be stationed back in the States. I would like to see him. Out here things are getting plenty hot. Tomorrow we make four strikes on the Philippines hiting airfields and Manila Bay for shiping. I still owe the government six dollars. Its been a long time since I've had money on the books. On our next pay day I'll start getting money on the books. Received four Christmas Cards. They are O.K. and have a good sketch of a Carrier.

Sunday/Monday November 19, 1944

Made three strikes and a fighter sweep. Opeation [sic]

*Schedule: 0620 VF Sweep Clark Field area. 0640 Strike 1-A shipping Subic Bay to Lingayen Gulf. 0900 Strike 1-B Aircraft & Aircraft Facilities Clark Field. 1200 Strike 1-C same as 1-B. For the day our **Hornet** planes damaged 70 Jap planes on the ground, only spotted one small ship and damaged it very heavily. It believed to have sunk. 1750 torpeadoe defense. Our fighters intercepted Jap planes and shot down 3 Jap planes. 1830 G.Q. was sounded, our ships were being attacked by Betties. One of our cruisers & battlewagon open fired and shot down one Bettie. The rest didn't come in close enough to attack.*

Monday/Tuesday November 20, 1944

*No strikes. We are heading for Ulithi to anchor and resupply. Found out names of ships of ours which were sunk on the Battle of the Philippine Sea. The Department identified six American ships lost in action including the light aircraft carrier **Princeton** previously announced sunk. The ships lost in addition to the **Princeton** were the escort carrier **Saint Lo** and **Glacier Bay**, the destroyers **Johnston** and **Hoel**, and the destroyer escort **Samuel B. Roberts.** A few lesser craft were also sunk. While we were anchored at Ulithi, the last time I know for a fact we had quite a few ships damaged.*

Tuesday/Wednesday November 21, 1944

Sent up a patrol 1430, came back 1730. Reed and I had five flats to fix. We have a machine in the shop and it works swell. At first it took us a little time to get use[d] to it, because there are so many parts which have to be change[d] and the job has to be done just right. We should reach Ulithi sometime tomorrow. Some of the cruisers & destroyers with our task group had practice. Some firing runs. Held field day. Movie, the name **Pittsburgh,** *starring John Wayne. The heat is toriffic [terrific]out here.*

Wednesday/Thursday November 22, 1944

Made a strike on Yap. Sent up 31 fighters all carrying fire bombs. Those new fire bombs will kill every thing within 90 to 150 feet. It has some kind of chemical which makes a hell of a fire which can't be put out. In our task group alone, 96 fire bombs were droped on Yap. G.Q. for entering port. 1400 anchored at Ulithi. Alot of mail came aboard most of which were Christmas cards. Held field day in shop. Fixed four tires. Its swell fixing tires with the new machine we have, and also we don't have to savalage [salvage] the old flat tires, tubes & busted wheels. All we do is throw them over the side.

Thursday/Friday November 23, 1944

I found out Nov 20 three Jap subs were sunk within the

sub nets of Ulithi. These Jap subs did manage to sink a tanker. Was put on a work party and I went on a supply ship to load supplies for the **Hornet.** *It was plenty of hard work, but we did get a few hours rest and we went over the side of the ship for a swim. We ate chow on this supply ship and it was lousy. Since it being Thanksgiving Day on our carrier for dinner they had turkey & Roast ham. I ended up eating some lousy roast beef. Got back to the ship 1730.*

Friday/Saturday November 24, 1944

Hornet News Flash. Further reports of the air strikes in and around Manila on 10 Nov. reveal the following total damage to shipping and installations by planes of the third Fleet: One Med. Cargoe Ship and one small Cargo Ship set fire in Subic Bay. Two medium Cargoe Ships burning and one small Coastal Cargoe ship sunk near San Fernando. One ship burned and another burning in Manila Bay. An oiler in flames and one medium cargoe ship, and 2 small oilers hit in Manila Bay. Five Luggers burning off Batangos, and another sunk at along[?]. A locomotive destroyed at Lucena. Our planes strafed a heavy cruiser near Santa Cruz, and many planes damaged on ground Del Carmen Fields & Molula.

Saturday/Sunday November 25, 1944

*Still anchored at Ulithi. Liberty party this morning, but I was on the duty section. Had a working party taking supplies below to store room #1 elevator. Taking on foul weather gear. I guess we are getting ready to go up north. What a surprise. Jocko is back with us. He aboard the **Hornet** tonight. He is a 3 star admiral now. Had a happy hour. Band played and boxing matches. Had torpedoe defense call. Bogies were picked up on the screen, but they turned out to be friendly. Taking on plenty of bombs, rockets, and ammunition. The **Enterprise** is going back to the States soon. **Wasp** is anchored right along side of us.*

Sunday/Monday November 26, 1944

*The happy hour we had last night was to celebrate the ship's ~~birthday~~ anniversary. By right the ship's anniversary is Nov. 29, but I guess maybe we will pull out before the 29 so that's why we celebrated it last night. Also the boxing contest with the **Enterprise** turned out to be a tie 3 to 3. Still taking on supplies. There are two hospital ships anchored here at Ulithi. Now that the **Enterprise** is going back to the States I don't know what ship will opporate [operate] with us, but with old (Jocko) Admiral Clark, the **Hornet** will be the flag ship in the task force again.*

Monday/Tuesday November 27, 1944

A task group pulled in from the Philippines. I found out the **Intrepet [Intrepid]** *was hit. She got hit with two suicide Jap dive bombers. She anchored on our starboard side. They say her flight deck is very badly damaged. There may be a few other ships that were hit, but I'm not sure of this. The Navy keeps most of this stuff confidential. The* **Enterprise** *left, and is on her way back to the States. Had a mail call. Some of the fellows received some of their mail badly burned. The Fleet Post at Eniwetok burned down. Sure wish none of my mail was burned.*

Tuesday/Wednesday November 28, 1944

The **Enterprise** *was called back, because of the* **Intrepet [Intrepid]** *being hit. I guess thats the reason. Had a mail call. I had Liberty, but I didn't go on the beach. These Islands we anchor [at] and have liberty are lousy. Guess once the Philippines belong to us then liberty will be O.K. Most of the fleet is anchored here. I guess we must have a few task groups standing by at the Philippines. We are waiting for meat supply, and vegetables. The supply ships are expected to be here tomorrow. Had a movie tonight* **Music in Manhattan.** *I wasn't on any work parties for the day.*

Wednesday/Thursday November 29, 1944

Our ship's anniversary. She is now one year old. I still can remember when I first came aboard the **Hornet** *Nov 28 [1943]. She was commissioned today [that day?] by Frank Knocks [Knox]. Went on a work party to a supply ship. On our way we stoped at the* **Wasp** *to pick up their work party. I got to talking to one of the boys from the* **Wasp.** *He told me they celebrated there [their] ship's anniversary Nov 24. We went to a supply ship. I had an easy job. All I had to do was to unhook the nets in the barge, then I went with the barge to the U.S.S.* **Yorktown.** *Last of all I went back to the ship with a barge loaded for the* **Hornet.**

Thursday/Friday November 30, 1944

Twelve of our bombers were taken off, and are going to be replaced by fighters. Altogether we are going to opporate [operate] with 72 fighters. Had mail call. Still having recreation on the hanger deck volley ball, basket ball & etc. Lt Koch & Lt Peters got transfered today. The **Intrepet** *is going back to Pearl Harbor, and then from there she may go back to the States. I found out the U.S.S.* **Cabot** *was hit while opporrating with the* **Intrepet** *at the Philippines. Had a movie on the hanger deck and the band played before the movie.*

'44 December

Friday/Saturday December 1, 1944

Left Ulithi 0715. While at sea our fighter replacements came aboard. Had a lot of flats. Had to fix 16 tires. Lousy day, plenty of rain. We were going to the Philippines, but I heard our orders were cancelled and that tomorrow we will go back to Ulithi an[d] anchor. The flag staff came on before leaving Ulithi this morning. Had zone inspection. Most of the day Reed & I worked fixing tires. Firing runs were cancelled. I don't know why our orders were cancelled but they must have had a good reason. Had mail call, and I received a few letters from home.

Saturday/Sunday December 2, 1944

Anchored 0715 at Ulithi again. The reason our orders were delayed was because of the weather conditions. They expect typhoons at sea. A Marine got hurt on the flight deck. A gun cleaner was taking care of the guns on a fighter plane and accident[ally] the guns went off. This Marine was the only one who was seriously hurt. He got hit in the leg. The doctor operated on him, and he is going to be alright. I had the mail censor duty. Reed and I fixed a TBF tire. A few fighters took off before entering port. They are going to land on the beach.

Sunday/Monday December 3, 1944

Had liberty but I didn't go on the beach. Its alot of trouble getting to the Island, and the shore is lousy for swimming. Went to mass this morning then I stayed on the flight deck for a few hours. Did some clothes washing & wrote a few letters. Miles made chief. Still anchored. I don't actually know how long we will stay in port. Recreation on the hanger deck football, basketball, volley ball & etc. Saw a movie the name **Mr. Winkle Goes to War** *starting Edward G. Robinson. It was a good picture.*

Monday/Tuesday December 4, 1944

Joe the baker died early this morning (0200). Joe yesterday noon was crushed between a barge & the ship's side. We [He?] was climbing the cargoe net & he slip[p]ed (liberty party). This morning I went on a work party to supply ship. I worked on the barge unloading nets. The **Essex** *was anchored on the port side of this supply ship & right along side of her was a repair ship. I found the* **Essex** *got hit at the Philippines the same ~~time~~ day the* **Intrepet [Intrepid]** *got hit. Altogether it was four carriers that got hit:* **Intrepet, Cabot, Independence,** *& the* **Essex.** *The* **Intrepet** *got the worst damage. She had to go back to the States for repairs.*

Tuesday/Wednesday December 5, 1944

No meat, just canned spam and vienna sausages. The meat supply ship hasn't reached Ulithi yet. Field day. The deck force has been working steady chiping paint. The armory next to ours has a recorder, and they have some good records. When I haven't anything to do then I listen to some recorders [records?]. Not much work to do in the shop, but I get on plenty of work parties. Saw a movie. The name of the picture was **Sunrise and Shine** *staring Linda Darnnet and Jack Oakie. It was a good picture.*

Wednesday/Thursday December 6, 1944

The meat supply ship still hasn't arrived here yet. Was on a work party. Went on a supply ship. We loaded up with canned pairs [pears], cut beets, & coffee. We got to this supply ship 0945, only loaded one barge & then went back to the ship. Still holding field day, and Zeke is working on number two elevator. They are making a new wire net. Held field day. Cleaned up the shop a little. Most of the planes are secured on the flight deck and the night fighters are standing by in case of emergency.

Thursday/Friday December 7, 1944

No fresh meat aboard ship. The meat supply ships haven't arrived at Ulithi yet. We have been having spam in place of meat. A lot of packages, Christmas ones, came aboard. So far I only received one package & that was from Shirley & Anthony. We now have 102 planes, 72 fighters and the other 30 are TBFs and SB2C. **Hornet** *Basketball team played the* **Handcock [Hancock].** *We won 41 to 9. The deck divisions have been working steady chipping paint over the side. I don't know how long we will be in port.*

Friday/Saturday December 8, 1944

Took on fresh meat, vegetables, & fresh fruit all night and day. I was put on a work party yesterday 1800 and I didn't finish work until 0700 this morning. We took a great heavy load of potatoes. I didn't have to do any work during the day, because I worked all night long, so I slept from 0800 this morning until 1630 this evening. I saw the movie. The name of the picture was **French Creek** *with Basil Rathbone. It was a good picture. Now that we have a full load of supplies, we should be leaving soon. 1915 [hours] torpeadoe defense.*

Saturday/Sunday December 9, 1944

Last night we had T.P.D. ~~It lasted about~~ *G.Q. was sounded and it lasted about a half hours [sic]. The ships were putting on steam just incase we should have been attacked. Planes took off the beach. Bogie was picked* up on the screen. Had a happy hour boxing match with the **Wasp.** We won 5 to 2. No movie. A tug boat pulled along side the starboard forward quarter and 150 new boots [sailors] were brought aboard. I hope we get a lot of these new seamen in our division, so they can transfer some of the rated men out of our div[ision].

Sunday/Monday December 10, 1944

Left Ulithi 0815, and now we are back out to sea again. Have plenty of planes aboard. Quite a few of our fighters met us out to sea. Altogether we have 72 fighters & between the TBFs and the SB2C amount to 30 which gives our carrier a total of 102 planes. Most of the day our fighters were practicing attacks on a skee sled aft of the fantail. They were shooting rockets & strafing the sled at high altitude dives, and boy they did some beautiful shooting. I could see those rockets skim through the air and hit darn close to the target. Had plenty of gunnery practice to[o] 5", 40 & 20 MM. Reed & I only fixed 4 tires. Two fighter crackups.

Monday/Tuesday December 11, 1944

Had the second patrol. Only had to fix three flats. Have a very big task force with us. Captain made a speech tonight. He told us about how large our task force escourt is. This time we [have] three of the Navy's best battlewagons, fourteen cruisers, and twenty destroyers. He was also telling us of the suicide Jap pilots which have been doing quite a bit of damage to our fleet ships. The Captain said with this task force if they come around he knows for sure our ship gunners can take care of them. He also told us safety rules to follow while be[ing] attacked. So far don't know our destination.

Tuesday/Wednesday December 12, 1944

The three battlewagons which are in our escourt are the **Iowa, Wisconsin,** *& the* **New Jersey.** *Three first line carriers* **Hornet, Lexington** *& the* **Handcock.** *Two converted carriers. One is the old night fighter carrier* **Independence.** *I don't know the name of the other one. Our destination is the Philippines again. We are going to make strikes at air fields in Luzon. We strike Dec. 14, 15, & 16. After that I don't know what we will do. Sent up a patrol. Quite a few flats. Altogether Reed & I fixed seven tires. With 72 fighters we ought to be having more tires to fix then we had in the past. Water was rough today.*

Wednesday/Thursday December 13, 1944

A bad rain storm delayed our patrol for a few hours. Tomorrow we strike Luzon. Our fighters are going to try & knock out very Jap plane they can possibly fine [find].

Our Captain made a speech. He said up to today our carrier since she has been commissioned has traveled ~~100~~ one hundred thousand miles. We sure have done a lot of moving. Already we could have gone around the world four times. The bombers & TBFs will stand by tomorrow. This operation is [to] try and get most of Jap planes into the air and then our fighters will take care of them. We may run into one of these suicide squadrons.

Thursday/Friday December 14, 1944

Fighters took off early this morning for Luzon. Very little enemy resistance and aircraft very few were spoted. For the day our **Hornet's** *planes shot down three Jap planes, 2 Valls & one Donna. They damaged about 10 planes on the ground. A huge Jap transport was spoted and our fighters strafed ~~it~~ & got a few ro[c]ket hits on it. They left this transport badly burning. The purpose of this operation was to knock out all enemy planes possible while our troops sneak in and hit the beach on Mindoro in the Philippines. Tomorrow the invasion of Mindoro takes place.*

Friday/Saturday December 15, 1944

MacArthur's troops hit the beach of Mindoro. Everything went along fine. No enemy resistance was accounted for. Our troops completely surprise[d] the Japs. They didn't figure Mindoro would have been our next invasion. Sent up a few strikes, and plenty of fighter patrols. Yesterday all the planes which went up were fighters. Spoted transport again & this time our planes sunk it. The purpose of our fighter patrols was to keep the Japs busy while our troops made a surprise invasion on Mindoro. The plan worked out fine.

Saturday/Sunday December 16, 1944

Today the 13,000 landing was made on the **Hornet.**
Sent up fighters, eight hops. Our night fighters took off 0230 this morning. These three day oporation [operation] we ran into very few Jap planes. The Japs must have a very weak air power. During the three days, the[y] only could send up thirteen planes to hit our task force. Our carriers' planes intercepted them & shot every one down. Also in our air group no pilot or crewman were lost. Six of our planes did hit the drink, but the pilots & crewmen were picked up. Our **Hornet** *planes shot down 2 Jap planes today.*

Sunday/Monday December 17, 1944

Early this morning complete task force refueled. Our Captain told us after we refuel we will stand by for further orders. We got our copies of the Hornet Tales and in it was an artical [article] about when we will go back to the States. It's been so long since we have been back to civilization & that's all we think of is going back. Well this artical kind of hurts. Our Captain said he is quite sure ~~our carrier~~ we won't be seeing the Golden Gates [sic] before June 16, 1945. Boy that's a long long way off. Water was plenty rough, had [and?] it was a tough job taking on fuel.

Monday/Tuesday December 18, 1944

Last night before lights out a Chicago war correspondent told us a little of his adventures. We also have a war correspondent that works for & Australian London paper. He was telling us about the Japs when they start com-[m]it[t]ing suicide when they know they are licked, and he said we are having the worst of all—a pilot in a plane ready to commit suicide is a very dangerous man. He told us how the Jap civilians on Saipan were commit-[t]ing suicide: razor, drowning, throwing hand grenades at each other & etc. Ran into a terrific storm today. This morning a plane broke loss [loose] on the **Monterey** [?] & caught on fire. It did plenty of damage. The water was so rough ~~it~~ a wave busted three of our roller curtains. We took all of gas out of [planes'] tanks.

Tuesday/Wednesday December 19, 1944

Storm cleared. Refueled destroyers. Last night before

lights out in the news broadcast they told us our old Captain Sample's CVEs were the escourt with this convoy of troops which landed on Mindoro. The convoy was attacked but every one of the forty six planes which made the attack were shot down by Admiral Sample's CVEs. Our loss was only one plane, and a destroyer of ours was slightly damage[d]. As for the troop convoy none of the troop ships were hit. As for w[h]ere we will go next I do not know yet.

Wednesday/Thursday December 20, 1944

*Early we sent up a search hop for three destroyers which must have turned over during the storm Dec. 18. One of the destroyers was the U.S.S. **Hall.** I don't know the names of the other two. It was a hell of a storm. Yesterday a search hop found ten surviers [survivors] in the water. I guess there must have been alot of men missing. Tomorrow we strike Luzon. Reed & I fixed seven flats. All the planes and objects on the ship have been secured to the decks. They believe we will hit another storm. I sure hope not. A destroyer out here was suppose to have mail for us. Sure wish it wasn't one which was lost at sea.*

Thursday/Friday December 21, 1944

Very early this morning one of our night fighters broke loose & fell over the side. All during the night we ran into a very bad storm. Because of the weather conditions all strikes over Luzon were cancelled. Hornet News Flash artical on our attacks on Luzon Dec 14, 15, & 16. Ships sunk: One large transport, three medium oilers, ten cargoe ships, two landing vessels, twelve small vessels. Ships damaged: four destroyers, two destroyer escourts, ten cargo vessels, twenty five small cargo vessels. Eight railroad trains and locomotives strafed & burned, & attacked a two hundred truck troop convoy near San Jose northeast of Manila & many other vehicles destroyed.

Friday/Saturday December 22, 1944

Left Philippines and we are going back to Ulithi. We will be anchored there during Christmas. I'm sure going to miss being home this Christmas. Its been a long time out here in the water far from civilization. If people could only see what some of these South Sea Islands look like, just sand barges with a few palm trees. A guy is better off to stay on the ship instead of going on some of these lousy Islands. Saipan looked like a swell Island, but none of the crew could get off the ship. They were still fighting it out with the Japs when we were anchored there. Had mail call. We got 22 bags from a D.E. this morning. Dad's birthday & I can't even buy him a present.

Saturday/Sunday December 23, 1944

On our way back to Ulithi. We will be there for Christmas and Christmas Eve. Early this morning a fighter plane captain fell off the flight deck. He was securing his plane at the aft tip end of the flight deck. He slip[p]ed & fell over. They droped a smoke bomb but when the destroyer got there they couldn't find no sign of him so he drowned. Had firing runs. Had mail call today & I received a few letters from the folks. Fixed four tires in the shop. Had one patrol, came back about 1650. Making good speed & will reach Ulithi tomorrow.

Sunday/Monday December 24, 1944

0945 we anchored at Ulithi and believe me plenty of mail came. All through the day we kept having mail calls. I received fourteen letters from the folks. Had a movie. The name of the picture was **Jam Session.** *It was a good movie. After the show we had a Happy Hour. Zuit [Zoot?] Suit & Red put on an act. This time Zuit Suit had a Zuit Suit an[d] Red was dressed like a girl. The band played an[d] they jitterbugged. It was really something to watch. After the show it was late so I waited for the Midnight mass. Also after the Happy Hour we had a mail call.*

Monday/Tuesday December 25, 1944

The Midnight Mass was a beautiful one. Along side of

the alter [altar] were eight of our Marines dressed in full dress uniform, four on each side of the alter. There was at least seven hundred men present and just about all of us received Holy Communion. After the mass I went to see Tony. He gave me a big hunk of turkey to make a sandwich with. It was 0215 when I hit the sack. Got up 0600. Received five packages & also a few Christmas cards & mail. The picture show we had **Destination Unknown** was lousy. Saw Tony, and he gave me some ice cream he got from a store keeper.

Tuesday/Wednesday December 26, 1944

Had liberty but I didn't go to the beach. I don't see why they call it liberty. All Ulithi is is a sand barge with a few palm trees. I would have gone swimming, but once was enough. They have a lousy beach to[o] many rocks & coral. I stayed in the shop an[d] wrote five letters. We had a mail call and I received a letter from home. Name of movie **Hail the Conquering Hero** cast Eddie Bracken and Ella Raines. It was a very good picture. Reed got stuck on a work party. Had to go on a supply ship. Am I glad I wasn't on the duty section today.

Wednesday/Thursday December 27, 1944

Was on a work party all day long, food supplies, can milk, can corn, & can peas & flour. We had to lower it *from the hanger deck to the supply compartments. On the gas bombs we have been using belly tanks. Now they are going to try and see how they will work with fins, which they use of 1000 pound bombs, welded on the end of belly tank. Some of the 2000 & 1000 pound bombs are being taken off because our SB2Cs are suppose to be replaced with F4U Corsairs. As for when we will take them on I do not know. Took on plenty of supplies. We may be staying out a long time when we leave Ulithi.*

Thursday/Friday December 28, 1944

In this morning's Hornet News Flash—between 16 and 25 Japanese planes raided the Airstrips on Saipan on Christmas Eve. Three of the planes were shot down by fighter planes and anti aircraft fire brought down another. One American plane was destroyed on the ground and several were damaged. It was not disclosed whether the destroyed or damaged planes were Superfortresses which are based on Saipan along with Liberators. Heavy bombers escourted by fighters hit Clark Field at Manila Christmas Day Manila time with forty-four tons of bombs in the sustained aerial offensive to knock out Japanese air bases on Luzon. American fighters in the Manila raid downed 39 of 50 Jap interceptors.

Friday/Saturday December 29, 1944

Went on a trash work party early this morning on an LST. We went about three miles outside the nets to empty trash over the side. Got back to the ship in time for chow. After chow was put on another work party. Only lasted about an hour & ½. Tomorrow we leave Ulithi and go back out to sea. I've got a feeling soon we will be seeing plenty of action, & with God's help I hope our fleet will have plenty of luck. Figure our fleet will be hit[t]ing Formosa. Boy what a hot spot that is. Saw a good movie **Since You Went Away.** *Ammunition was being taken aboard all day long. Rain & cloudy most of the day.*

Saturday/Sunday December 30, 1944

0915 left Ulithi. I was out on the for[e]castle watching one of our battle ships past close by. Have a big task group. Three carriers **Lexington, Handcock,** *& us. Also two converted carriers,* a few *two cruisers & twenty destroyers. Had plenty of gunnery practice. Our 5" did very well in shooting down target sleaves out of the sky. Had two patrols 1st & fourth. Reed & I had to get some new tires. We fixed three tires. Our store rooms were loaded while anchored at Ulithi. We may be staying out along [sic] time, and I wouldn't be a bit surprise[d] if we end up in the China Seas.*

Sunday/Monday December 31, 1944

Had the 2, 3, & 4th patrol. So far our Captain hasn't told us our destination. Well, Diary, tomorrow starts another year. In this past year of 1944 I only had two liberties ashore, one at San Diego & the other was pearl Harbor. When we left the States we went through the Panama Canal, up to San Diego, Pearl Harbor & then to Majuro. When we had our shake down cruise we went up around Bermuda. Shake down only lasted about two weeks. From Nov 28, 1944 to Feb [blank space] 1944 stayed back in the States. Went to mass this morning. Received two packages. They took four months to reach me. Storm at Cape Paters. Almost fell over the side while getting wind breakers.

'45 January

Monday/Tuesday January 1, 1945

Today starts off a new year. Had air department reveille 0215. This morning sent off 30 fighters, 0600 Planes did practicing dives on our task group. Gunnery department had firing runs most of the day, 5", 40mm, 20mm. Fixed two tires. Started to make a knife blade. Hope it comes out good. Chow is lousy. Most of the food is old, and at times you spot quite a few bugs in your chow. Still don't know our destination. Went to mass 1400. I figure we will be hitting Formosa. Luzon should be invaded soon once MacArthur gets the Philippines squared away, then it will be plenty hot for the Japs.

Tuesday/Wednesday January 2, 1945

Last night just before lights out the skipper wish[ed] us a Happy New Years, and gave us the straight dope. We strike Formosa. Early this morning we refueled. It was plenty cold and quite a surprise to all of us. Some of the fellows on the flight deck had on sweaters or jackets. Water plenty rough. Had a few patrols. Four flats had to fix. We had a big escourt with us this time. The last time we were at Formosa, we had a few destroyers & two cruisers. This time we even have three battle wagons with us. Well tomorrow we will see if the Japs will put on the heat like the last time.

Wednesday/Thursday January 3, 1945

Made strikes on Formosa. A convoy was sighted anchored and also another convoy was spoted well on its way. Our **Hornet's** *planes hit this convoy which was at sea. It consisted of 3 large troop ships, with a destroyer escort. Our bombers got 3 1000 lb. hits on one of the troop ships & left it badly burning. Weather was very bad. It was so cloudy over the target that after the 3rd strike, rest of flights were canceled. Our fighters early this morning shot down one Betty. Our planes also bombed air fields at Formosa. Its very cold. Formosa is only 90 miles from the China Coast. One TBF went in the drink, but crew got out, and were picked up by a destroyer.*

Thursday/Friday January 4, 1945

Weather was very bad. Awful cloudy, & rain. Sent up a few search hops for the convoy which our planes attacked yesterday. Also sent a few strikes over Formosa. Pilots had to fly by radio because of the very bad weather conditions. No sight of the convoy was found, and when the Captain asked the bomber Comdr. what damage he did over the target, he said I don't know but I think we did bomb Formosa. It was very cloudy. Most of the strikes were canceled. One fighter number 7 crashed on the flight deck. Reed & I fixed 9 tires. Tomorrow we will refuel.

Friday/Saturday January 5, 1945

Early this morning the rest of the fleet was in sight. While we refueled from tankers, the water was very calm & it is getting plenty warm again, because we are heading back to the Philippines. Had the 3rd patrol. A destroyer came along side & mail was taken aboard. It wasn't much, and there was no letter for me. Captain made a speech tonight. He said tomorrow we will make strikes on Luzon and at the same time our Army planes from Leyte will be hitting Manila area. The Navy and Army Air Corps will strike together. Strikes will only last one day. After that, stand by for further orders.

Saturday/Sunday January 6, 1945

Made strikes on Luzon. Weather conditions over target were very bad. Our fighters ran into five Zekes. They shot two down & 2 probably shot down. On the strikes our planes destroyed about 30 Jap trucks & a few trains. The Japs are getting ready on Luzon. Purpose of strikes were to wipe out all of Jap air power seen. Also damage all planes seen on airfields. Early this morning a search hop was sent up for Jap shiping. Night fighter crashed. His landing gear would not release. We changed our course and now heading for Formosa. So far in the last few days weather conditions have been very bad.

Sunday/Monday January 7, 1945

Was on our way to Formosa but were called back. Sent up strikes all day long over Luzon. The reason we were called back was because the 7th fleet which is softening spots on Luzon for the Invasion is running into alot of trouble with these Jap suicide pilots. Just early this morning 12 Jap suicide planes crashed into 12 different ships in the seventh fleet. Yesterday these suicide pilots sunk a CVE in the 7th fleet. Our planes bombed airfields on Luzon. They shot out of the sky **Hornet's** *planes 2 Tojos. One of our fighter pilots and a few P-51s ran into a few Tojos and another pilot of ours got an Oscar. Bad day for us. We lost 3 pilots, and one radioman.*

Monday/Tuesday January 8, 1945

Fleet refueled. Standing by at the Philippines. Over the speaker on the news broadcast the air group 2 fighter pilot which made a force[d] landing on Leyte had alot to do with MacArthur's change in the Invasion plans. In Hornet's News Flash damage our fleet did at Formosa Jan 2 & 3: Apparently sunk or destroyed 1 Large Cargoe Ship West of Takoo, 1 Med Cargoe Ship at Keelung, 1 Small Cargoe Ship at Keelung, 1 Patrol Craft, 10 Small Coastal Cargoe Ships, 11 small craft. Damaged 1 Destroyer, 3 Destroyer Escorts, 6 Patrol Craft, 1 Landing

Ship, 2 Landing Craft, 2 Large Cargoe Ships, 1 Med Cargoe Ship, 34 Small Cargoe Ships, 7 small craft. On ground destroyed 111 enemy air craft.

Tuesday/Wednesday January 9, 1945

*Invasion of Luzon took place. We covered the invasion by making strikes on Formosa. Had torpedoe defense twice. **Yorktown** patrol splashed one Jap plane. Our air group did some good hiting on Jap shipping. They sunk two heavy freighters. Also did heavy damage to 11 ships, one of which was a destroyer. At Formosa, they destroyed 15 Jap planes on the field. As yet no reports of the invasion was given. Our air group also destroyed 5 locomotives. We lost one TBF. Two of the crew were seen in a raft. Also a radioman in one of our TBFs was killed. Captain told us tonight that tomorrow we will be opporating in the South China Seas.*

Wednesday/Thursday January 10, 1945

Last night our fleet went through the channel between Luzon & Formosa into the South China Sea. Already we sent a few patrols over to the China Coast. Early this morning we had G.Q. I saw a Jap Betty eight miles away from our carrier shot down by a night fighter on patrol. Our skipper tells us we ought to run into plenty of action out here. Reports on Invasion: landed 0930 Tuesday Jan

9, 1945 on Luzon Island, Lingayen Gulf, and secured four beachheads against light opposition. Troop convoys were attacked. The 7th fleet shot down 79 Jap planes, sunk two destroyers, a sub & a cargoe. Some naval losses were suffered, but they did not stop the invasion.

Thursday/Friday January 11, 1945

Tomorrow we make strikes on ship[p]ing & airfields on the French Indo China coast. This afternoon the **Ticonderoga's** *patrol shot down three Jap seaplanes 25 miles from the ship. Captain told us tonight one of our fleet patrols spotted two Jap battle wagons, four cruisers, and quite a few destroyers anchored at the same Gulf on the China Coast. We are going after these major fleet ships of Japan. We make a speed run. Captain said so far the fleet believes the Japs haven't spotted us, but those planes the* **Ticonderoga** *shot down may have spoted our fleet and gave us away.*

Friday/Saturday January 12, 1945

Our first strikes went after the major fleet ships of Japan. When our planes got there they found these fleet ships gone so the Japs must have suspected us. But our planes did very well early this morning. Our task force attacked a Jap convoy: 4 DEs, 1 Heavy Freighter & three large cargoe ships. All were sunk except one DE which was left badly burning. Our **Hornet's** *planes scored on this convoy four bomb hits & three torpedoe hits. Also attacked another convoy. Our* **Hornet's** *planes sunk a light cruiser, got three bomb hits & two torpedoe hits. All attacks on shipping made was centered between Saigon and Camrahn Bay. We lost a TBF pilot & crew. Also a fighter [made a] force landing in a rice field. Natives were seen taking him [to] cover. Ship was only 45 miles from the Coast of French Indo China. Also* **Hornet's** *planes destroyed 14 planes on Jap fields.*

Saturday/Sunday January 13, 1945

Jan 13, 1945. Bad storm hit. A typhoon. Even though the water was very rough some of our ships refueled. When we came through the channel between Formosa & Luzon a convoy of tankers came with us. On the hurricane our fleet hit Dec 18, we lost three destroyers: **Hull, Spence,** *&* **Monaghan.** *Also seven other craft due to enemy action. The Navy Department have already announced the losses. Was in flight quarters. Planes were being respoted, Because of the bad weather conditions strikes were canceled. Tomorrow our carrier will refuel.*

Sunday/Monday January 14, 1945

Getting very cold. Have to wear a jacket, & in the compartment instead of having trouble sleeping with the heat its the other way around. Now we have trouble sleeping with it being so cold. Have to sleep with blankets now. Very cloudy & windy. Refueled from tanker. Only had one flat to fix. We did send up a long search hop. They covered over four hundred miles. Our planes spotted a lon[e]ly Jap freighter & they sunk it. Had torpedoe defense. One of our gunners died from a heart attack while going to his G.Q. station. Tomorrow we make strikes at Hong Kong, Formosa shipping & airfields.

Monday/Tuesday January 15, 1945

*Strikes from Hong Kong to Southern tip of Formosa. Weather conditions very poor. Awful cloudy. Didn't run into much air opposition. Our **Hornet** planes shot down four Zekes. Also got three bomb hits on a destroyer. It believe[d] to have sunk. Damaged quite a few enemy planes on Jap air fields. A few strikes were cancelled because of the very poor weather conditions. Had torpedoe defense & were in condition one easy most of day. Also our planes beached a Jap destroyer by getting several rocket hits. Our loss: one bomber pilot was picked up but the gunner went down. Tomorrow all strikes on Hong Kong.*

Tuesday/Wednesday January 16, 1945

All strikes made on Hong Kong. Weather conditions were

*very bad. Our **Hornet** planes did plenty of damage over Hong Kong. They completely destroyed 3 large docks & one heavy freighter which was tied to the dock was blasted to hell. On shipping they heav[i]ly damaged eight in the bay, did heavy damage on airfields & factories. Also destroyed one large oil dump & left it badly burning. Last strike ran into alot of anti aircraft fire. Hit a convoy with a large Jap DE screen which sent up a heavy fire. Lost two fighters & One TBF. Radioed position of one of the fighter pilots to sub.*

Wednesday/Thursday January 17, 1945

Was suppose to refuel but water was very rough. Took on replacement planes. Have quite a few planes being fixed on the hanger deck. Alot of them have been shot up over the targets. Water very rough. Expect another storm. Reed & I fixed three tires. Sent up a search hop, and also had a few patrols. This last few weeks weather conditions have been very bad. So far our fleet hasn't had much trouble with Jap air power. Here we are in the South China Seas and still the Japs won't send in planes. If they do its always a very few planes. Don't know what we will do tomorrow.

Thursday/Friday January 18, 1945

Water still very rough. Run into a bad storm. Had Flight

Quarters. 2nd and 4th patrol. Reed & I had 11 flats to fix. Weather conditions lately have been very bad. What our next strike will be as yet I do not know. When the planes came in for landings, the pressure burst of[f] the tires on two fighters. Blew the fairings off. Still in a bad typhoon, but yet our fleet sends up patrols. Still only one slice of bread for each meal. A lot of flour spoiled. This noon's chow I found a live bug crawling around my tray. I don't mind them when they are dead, but this one was walking over the potatoes & then jumped into the peas.

Friday/Saturday January 19, 1945

Still out in the China Sea. It's really a surprise the Japs so far haven't put up enough aircraft to attack our fleet. Whenever they do send up planes it's always a very few, but they are plenty dangerous. These banzai pilots have been causing us plenty of trouble. I saw some of that Banzai bunch when we were at Formosa, and they did plenty of damage. Sent up two patrols. Refueled early this morning. It's a beautiful day. The sun is shining for a change, and the water is alot calmer than it has been in the last few days.

Saturday/Sunday January 20, 1945

Captain told us our fleet would be going out of the straights [straits] between Luzon & Formosa. A Jap plane was shot down & so the Japs found out we are trying to leave the S. China Sea, and we did run into plenty of trouble. G.Q. sounded 1645 & lasted [until] 2210. I was out on the fantail during G.Q. I saw our ships in full speed heading for the straights. Our patrols shot down 12 Jap planes, but still bogies approached. Our ships open fired. It was dark. Saw huge flashes & tracers. Our guns open fired once. 1100 I saw land portside. It was Island of the Formosa group. I don't think the Japs damaged any of our ships while going through the straights.

Sunday/Monday January 21, 1945

Made strikes on Formosa. Jap planes came in. I saw 38.3 under attack. Jap Banzais did alot of damage. Two bomb hits on the **Ticonderoga.** *Captain & Ex. badly injured. A bomb hit on the U.S.S.* **Langley** *& a bomb hit on a destroyer. 43 Jap planes shot down over 38.3. Some of our* **Hornet's** *planes got in the dog fights. They shot down 3 Jap planes. While 38.3 was under attack, the carrier on our starboard side U.S.S.* **Handcock** *[Hancock] planes were landing. I happen[ed] to be watching. I saw a TBF land and saw it exploed [explode] about midship. The cause: a five hundred pound bomb went off. Saw a heavy fire spread on the* **Handcock's** *flight deck. Bad day. Lost alot of lives.*

Monday/Tuesday January 22, 1945

Our carrier planes yesterday did do alot of damage to the Japs. 43 Jap planes shot down while 38.3 was being attacked. Over 100 planes destroyed on Formosa airfields & seven cargoe ships sunk. Today we made strikes on the Nansei Shoto Group only about two hundred miles away from Japan. Our planes didn't run into Jap aircraft & very few planes were spoted on airfields there. No shiping was spotted. We are back in the Pacific again. I almost forgot yesterday Jan 21 G.Q. sounded 1300 & lasted until 2200. I saw ships on our port side open fire, and saw a Jap plane burst into flames.

Tuesday/Wednesday January 23, 1945

Fleet refueled but our carrier refueled from tanker 1700. Things qui[e]ted down for awhile. The Japs kept us going. Our Captain told us we are heading towards the Philippines & will stand by for orders. Painting the shop. No planes took off. Church services for Sunday were held this morning out on the forcastle. On this opporation quite a few of our ships were hit and alot of lives must have been lost. These Japs send in at times a few planes but they sure do alot of damage these Banzai pilots. Suppose to take on mail soon.

Wednesday/Thursday January 24, 1945

Having beautiful weather again. Out here it's really something to see some of these beautiful sunsets. We're painting the shop. No more strikes. We are heading back to port. Destroyer came along side. At last we got mail aboard, and it sure was good to hear from the folks again. Had firing runs 40mm, 20mm, & 5". Lousy chow. It's going to be a happy day for me when this war is over. I'm getting sick of seeing just water every day. Had the 2nd & fourth patrol.

Thursday/Friday January 25, 1945

Had another mail call, and I got a letter from Anthony, also a few pictures of my niece. She sure is nice looking. Still working with the paint work in the shop. Some of our planes took off and are going to land on airfield Ulithi. Tomorrow we anchor at Ulithi. Some of these small Islands aren't very much to look at. One good thing we will receive mail. Had a movie. Name of picture **New Yorktown.** *It was a good picture.*

Friday/Saturday January 26, 1945

Anchored at Ulithi 0930. Lousy day. Plenty of rain. Working parties started. Plenty of bombs coming aboard. Liberty parties. Most of the fellows don't go over to the beach because there isn't much to do, and its lousy for swimming. Movie **American Romance.** *It was a good picture.*

Saturday/Sunday January 27, 1945

Commissary. Lot of food supplies coming on. Found out the **Ticonderoga** *was hit very bad. Two banzai planes & one bomb hit. They weren't even in G.Q. when these Japs made their suicide dive. Over two hundred casualties, 110 of them were killed. On the* **Handcock** *that five hundred pound bomb did plenty of damage: 53 killed, 101 wounded. I saw a Jeep carrier on our port side. She looks like she was hit. Her flight deck is at an odd angle.*

Sunday/Monday January 28, 1945

Went to mass early this morning. Ammunition still coming aboard. ~~Some~~ *We took on some of the* **Ticonderoga's** *planes for replacements. Our Air Group #11 is getting off before we go back out to sea. Air Group #12 is coming aboard. That means we are good for another four months out here before going back to the States. Air Group aboard our ship got 102 Jap planes to their credit. As for shipping, I don't know the exact score. Altogether our ship has 374 Jap planes to its credit.*

Monday/Tuesday January 29, 1945

Found out we will be staying in port for quite some time. Plenty of supplies coming aboard. Found out [about] that Jeep carrier which is at an odd angle. It didn't get hit. The name of the carrier is the **Independence.** *They just filled her tanks on the port side so that they could do some repair work below the water line on the starboard side. Had a movie on the hanger deck. Some of our fellows went over to the* **Enterprise** *for a boxing match.*

Tuesday/Wednesday January 30, 1945

Went on a work party handling ship stores. Tanker came along side and we took on fuel. I found out our air Group 11 is leaving us, and Air Group #17 is coming aboard. The air group is packing up and they should be leaving the ship soon. Had a movie on the hanger deck. Nothing of any importance happen[ed] today.

Wednesday/Thursday January 31, 1945

Went on Liberty this morning Ulithi. Name of Island was Mog Mog Island. Had a beer party. Three cans of beer per man. Left ship 0730 and came back 1330. Still plenty of work parties taking on alot of bombs. On our way back from Mog Mog we pass[ed] the U.S.S. **Handcock.** *They had a repair ship along side of here [her]. The* **Ticonderoga** *has already left for the States. She was* *hit very bad. 110 of her men were buried at sea. Air Group 11 leaves the ship tomorrow and air group 17 comes on.*

'45 February

Thursday/Friday February 1, 1945

Air group left the ship. The band played while they left on barges & landing crafts. During hanger deck parade, medal awards were presented to members of Air Group 11 Squadron: Air Medal, Navy Cross, Purple Heart, Bronze Star & etc. The new air group came on tonight. Boy what a mess. They were all plastered, and what a rug[g]ed bunch of looking fellow[s]. If they fly like they look I pity the Japs. Now that we have a new air group it means another good four months at sea.

Friday/Saturday February 2, 1945

*Air Group 17 brought and [sic] awful lot of equipment aboard with them. I found out Air Group 17 was on the U.S.S. **Bunker Hill.** Went back to the States, had a leave, and the air group was reorganized. The fighter squadron is a new one. Most of Air Group 11 old bunch in the Torpeado & Bomber Squadron are still with the air group. Well soon we will find out how hot this Air Group is. New insignia from [illegible] for our planes. It's no more a circle. It's two with squares like this on tail [picture drawn].*

Saturday/Sunday February 3, 1945

Bombs are still coming aboard. Tomorrow we are going out to sea to break in Air Group 17. We are going to stay out about three days, then we are coming back to Ulithi. I found out a fellow who went through boot camp with me is an Electrician Mate 3rd with the night fighter squadron. I didn't get a chance to see him yet but Gostin & Zeke already saw him. Had a wat[c]h on flight deck 20 2400 [hours]. Supplies came aboard tonight right after the movies, and work parties worked all night long.

Sunday/Monday February 4, 1945

Left Ulithi 0745. Had G.Q. while leaving the Island, three full deck loads took off. Altogether we had four launches. The last one was the 4th patrol. Reed and I fixed six tires. The air group did very well. I watched most of them take off the **Hornet** *for the first time. For practice instead of water bombs they used 100 pound bombs, rockets, and gunnery practice on a skee slead about three hundred yards away from the ship. It was really something to watch. Had ship gunnery practice, also night practice 5", 40 & 20mm. I think the Air Group is going to be tops.*

Monday/Tuesday February 5, 1945

Flight Quarters all day. Ships guns had firing runs. Had practice attacks on the ship. Our pilots dived at a skee slead we were towing and dropped one [hundred?] pound bombs, rockets and gunnery practice. One bomb landed so close to the carrier that one of the fellows found a hunk of scrap metal on the flight deck. Two of our planes collided, a fighter & bomber. The crew were killed. One fighter made an emergency landing, and was picked up by a destroyer. **Wasp** *with us breaking in a Marine squadron of Corsairs.*

Tuesday/Wednesday February 6, 1945

Fixed eight tires. Altogether in two in [and?] and a half days operation we fixed twenty four tires. Watched our planes making practice attacks and droping 100 pound bombs & rockets at the skee slead we were towing. Some of those 100 pound bombs landed plenty close to the ship. ~~One of the fellows picked up a bunch of scrap metal off the flight deck~~ *Went back to port, anchored at Ulithi 1600. While we were coming into port I saw a bunch of troop ships with an escourt leaving Ulithi. Had a movie on the hanger deck.*

Wednesday/Thursday February 7, 1945

Early this morning I saw four carriers, a heavy cruiser, & two Jeep carriers pass through the nets. These four carriers are going to join us on our next operation. Some how I got a feeling we are going to hit Tokyo or some place in Japan. Our air Group 17 looks like a hot bunch, soon we will find out. It ought to be plenty hot for us. The carriers four that came in are the U.S.S. **Saratoga,**

Bunker Hill, *& two brand new ones,* **Bennington & Randolph.** *Plenty of bombs & Incindorary [incendiary] bombs came aboard.*

Thursday/Friday February 8, 1945

Had a few mail calls. The old **Saratoga** *is anchored along side of us. Altogether I think we will be oporating with twelve ~~carriers~~ first line carriers. While breaking in our new Air Group 17 out to sea the* **Wasp** *was with us. She has a Corsair Squadron of Marine pilots. They were breaking them in as part of their Air Group. MacArthur is doing very well in the Philippines. Already he has taken Manila the Capitol city of the Philippines. I don't think we will stay in port long, and soon we will be making strikes.*

Friday/Saturday February 9, 1945

Last Night we saw the movie **The Fighting Lady.** *It was pictures taken actually of our carriers. Most of the pictures were taken on the U.S.S.* **Yorktown** *when our Admiral Clark was Captain on the U.S.S.* **Yorktown.** *Before the picture started the Lt who was assigned to the U.S.S.* **Bunker Hill** *filmed the picture. If it wasn't for our Admiral he said the film would never have been made. Had trouble getting equipment & taking pictures of going through the [Panama] Cannal [Canal], but he said our Admiral straighten[ed] every thing out. Tonight the movie*

we had: **To Have or To Have Not** *[starring]*
Humphrey Bogart.

Saturday/Sunday February 10, 1945

1045 we pulled out of Ulithi. Back out to sea again.
So far I don't know our destination, but it looks like
things will be plenty hot soon. The Bonins are going to
be invaded, and somehow we all got a feeling this time
it will be Tokyo. Our ship's guns had firing runs. Took
some planes from the beach. They landed while out
to sea. Jocko is in Command of our task group, so we
are the flagship again. In our task group three first line
carriers: **Hornet, Bennington,** *and the* **Wasp.**

Sunday/Monday February 11, 1945

Our ship's guns had gunnery practice. Also our planes
did practicing on a skee slead towed by the U.S.S.
Bennington. *This morning I saw the jeep carrier*
Bellawoods *on fire. A plane crashed and made a*
hell of a fire. It burned clean through the flight deck.
I saw flames drop through the flight deck on to the
hanger deck, but the **Bellawoods** *is still with us.*
Have a another war correspondent, Mr. Smith, from
the **Chicago Tribune.** *Also found out one of the*
president's son[s] is with our flag staff, and aboard our
ship.

Monday/Tuesday February 12, 1945

Planes did alot of practice. I saw practice runs our task force planes made on a skee slead towed by the **Wasp.** *It was really something to see the way our planes would come down in a dive. Also our ships guns had firing runs. Today we were notified we will be receiving heavy winter underwear, woolen socks and foul weather gear. At last its Tokyo or Yokohama. This ought to be a hot time for Japan. We're heading full speed towards Japan, don't know yet when we strike.*

Tuesday/Wednesday February 13, 1945

Foul weather gear is being issued out. Had two patrols, and refueled from tanker. Stretchers plenty of them are being put in compartments on the ship. Its going to be Japan this time. So far the Captain hasn't told us our destination, but it's very easy to guess. Ships guns had gunnery practice through out most of the day. I finished making my knife. Held pay day on the hanger deck. In a few more days it should be plenty cold, and in another way it should be plenty hot.

Wednesday/Thursday February 14, 1945

At last the Captain gave us the straight dope. Feb 16 we will make strikes on Tokyo. The Captain said he believes our ships is well prepared to hit Tokyo. Had first aid lectures. This time they expect alot of casualties. The underwear we are getting are very heavy winter underwear, and the socks are heavy woolen ones. The **Bennington** *had a very bad crash landing on their carrier. For a minute we thought we may have had to take the rest of her planes before it got to[o] dark, but they managed to get straighten[ed] out.*

Thursday/Friday February 15, 1945

The weather really changed. Its plenty cold and the water is very rough. One of our patrols spotted a Jap picket boat (patrol boat) 40 miles away from our task force. They strafed the boat and a cruiser of ours went in close enough to sink it. We had the fourth patrol. It was really a sight to see all the guys wearing long winter underwear and very heavy woolen socks. Our Captain made a speech. He told us tomorrow will be remembered as a great day of history. Well it looks like plenty of fire works tomorrow. Tokyo.

Friday/Saturday February 16, 1945

G.Q. sounded 0600. Made strikes on Tokio [sic] all day. We stayed in our General Quarter stations. Tokio didn't turn out to be such a hot spot at all. A few planes were shot down by some of our ships. I saw one burning out over the horizon. Believe me, it was plenty cold today.

Had a couple of crash landings. Three of our planes went in the drink. Most of the crew were picked up. Our ship went in as close as 86 miles from Tokio. For the day the fleet carrier's planes shot down 256 Jap planes. A picket boat sunk 12 miles away from our ship.

Saturday/Sunday February 17, 1945

Our fighter squadron sunk a Jap CVE carrier. Weather conditions very bad, and plenty cold. Today our planes droped propaganda pamphlets over Tokio. I got a couple of them. Our fighters as up to now I know shot down three Jap planes, but as for two day strike on Tokio I do not know their complete score. Our Captain made a speech. He said he was proud of our ship and its crew, and that each man should be proud to be on the first carrier strike on Tokio. Also said from now on Tokio will be hit plenty. Also found out some reels of **The Fighting Lady** *droped over target.*

Sunday/Monday February 18, 1945

On our way back to the Bonin we made a strike on the Bonin Islands. 1300 to 1840. The Bonin Island will be invaded by our troops soon. Had G.Q. tonight 1810, bogies on the screen. The Captain told us we are opporating between the Bonins and Japan. Its still cold. So far we haven't run into much Jap air power at all. Its a surprise the Japs didn't send up huge amounts of planes when we made strikes on their home land. It looks like the Japs lack air power. Our planes are really giving them hell.

Monday/Tuesday February 19, 1945

Refueled from tanker. Took on ammunition off an ammunition ship while at sea. It was really something to watch nets being transfered back an[d] forth. The nets were made of very strong steel cables. Took on quite a few 2000 pound bombs. Ships had gunnery practice. Over the speaker system it was announced that the Bonins were invaded Island Iwo Jima. Part of our fleet is covering the invasion. We refueled today. Tomorrow our part of the fleet will go in an[d] cover the Bonin invasion, and the other half will leave, come out to sea and refuel. Over our ship's radio we could hear our carrier pilots radioing in reports.

Tuesday/Wednesday February 20, 1945

Covering the Bonin Invasion. Boogies picked up on the screen but G.Q. wasn't sounded. Had a few strikes launched to hit certain areas on the Bonins. I don't think we will stay on the Bonins long. Soon we will be going back to Tokyo, and bomb the hell out of Japan. If Japs are going to really come out an[d] hit our fleet now is their chance. Today it wasn't to[o] cold but if we go back to Japan I'll have to wear those darn woolen underwear. A few of our planes were shot up badly.

Wednesday/Thursday February 21, 1945

The place Iwo Jima is in Volcano Island group. Sent up three strikes. Are [Our] Marine troops are meeting stiff resistance. Boogies on screen G.Q. was sounded 1600. Are [Our] task force about 1900 started to fire at two Jap planes which kept flying over task force. I was out on the fantail watching the fire works. Every once in awhile some of our ships would open fire. Sent up night fighters. About 2100 the screen was cleared and G.Q. was secured. We are still close to Japan. Its a surprise the Japs don't try to make a real heavy raid on our fleet.

Thursday/Friday February 22, 1945

Scuttlebutt going around about our fleet being attacked. They say the **Saratoga** *took four fish and also a few other ships in same task force with the* **Saratoga** *were hit. They were attacked last night. I wouldn't be surprised if the* **Saratoga** *did get hit. Since we have been out here quite a few of our first line carriers have been damaged badly. Also a few Jeep carriers have been sunk. So far our* **Hornet** *has been very lucky. We came darn close to getting hit at Formosa especially. Soon we will be going* *back to Tokyo. Made a few strikes on Iwo Jima.*

Friday/Saturday February 23, 1945

*Made strikes on Iwo Jima. Our Marines are having
a tough fight and an awful lot of casualties. Iwo Jima
turned out to be one of the worst battles in the Pacific.
Already our casualties listed are close to 4500. The
Japs are duged [dug] into caves in mountains and our
Marines can only advance by yards. Our planes go in
and straf[e], bomb & shoot rockets. Army and Navy
planes bombed the hell out of that small Island Iwo
Jima. Also our battle wagons, cruisers & destroyers
shelled Iwo Jima many times.*

Saturday/Sunday February 24, 1945

*Bad storm last night. We're not up around the Bonins
now, we are going up to Japan tomorrow. We are going
to make strikes on large naval base in Japan. So far they
have [not?] told us our destination. It's just scuttlebutt
going around, but tomorrow for sure I will know the
name of this naval base. In this morning['s] news flash
it mentioned some of our ships got hit near Iwo Jima. 21
that was the night we were under attack. It looks like the*
Saratoga *and maybe a few more of our ships got hit.*

Sunday/Monday February 25, 1945

Made strikes on Japan, Tokyo and a Jap naval base was the schedule. Weather conditions turned out very bad. Strikes were cancelled 1400. Pilots claim visibility over target was very poor. Water very rough. Some of our pilots did sho[o]t down a few Jap planes, but I don't know the exact number. The last strike which landed droped their bombs in the water. Pilots said they couldn't see [as] ceiling was very poor. I guess we will be making more strikes on Japan tomorrow. Sure wish the weather will change to the best.

Monday/Tuesday February 26, 1945

Bad storm last night. Still cold out here around Japan. I was on watch from 0400 to 0800. Bad storm. I saw a wave go right over the flight deck. Good thing I wasn't in that wave's way or I would have gone over the side. Wind was very strong. All strikes were canceled today because of the storm. I worked on my knife. Its all finished, case and knife. Had G.Q. send up fighters. One Jap plane was shot down. Also send up a fighter sweep over Nansei Shoto I believe it was.

Epilogue

The diary ends. From mid-April until the end of May, the *Hornet* helped support land forces during construction of airfields on Okinawa. On June 5, 1945, her flight deck overhang was severely damaged by a typhoon's 100-foot waves. On June 19, the *Hornet* and her sailors left the Leyte Gulf, finally beginning their westward journey home to the States. After over 18 months at sea, they docked in Alameda, California, on July 8, 1945. The *Hornet* would see active duty until 1970—including the Vietnam War, and recovering both the Apollo 11 and Apollo 12 space capsules.[2] In 2008, she is a floating museum, docked once again in Alameda, California, welcoming visitors year around (www.uss-hornet.org).

During his service aboard the *Hornet*, Carmen earned the American Theater Medal, Asiatic-Pacific Medal with 9 Stars, Philippine Liberation Medal with 2 Stars, and the World War II Victory Medal. He was honorably discharged from the USN Personnel Separation Center, Lido Beach, Long Island, New York, on February 18, 1946.

Appendix A

International Dateline Ceremony

"Passed the Equator today. That means I'm a shell back instead of a polly wog. Also we are a day ahead. By rights its Fri. but out here its Sat. I'm going by old time so I won't have to skip a [diary] page. . . . Skipped a day because we passed the National [International] Date Line."

     –March 24, 1944

"Inniciation [initiation] from a pollywog to a shellback will be given tomorrow."

     –May 2, 1944

"Today I became a shellback and what a beauty of an Inniciation. We had to go through the middle, dumped in a tub of salt water, and what a haircut they gave me. I'm practically bald."

     –May 3, 1944

The Captain welcomes King Neptunus Rex aboard.

Sail ho! Where away? Far away!

The Royal Works.

Getting the Hot (ahem) Foot.

TBF Pilot on Lookout Watch.

Garrison Finish!

The long and short of it.

Where's Earl Carroll?

Looking for a short beer.

Appendix B

Interview with Barbara Franceschelli

July 20, 2005
New Britain, Connecticut

Naomi: I am so glad that we could get together after all these years. I am trying to contact all the people that once worked at my father's coat factory and people that our family knew.

Barbara: I am happy if I can help in any way. Yes, Mom sewed for your father's factory six days a week, from 7 am to 6 pm as all the factories did in those [war] days. You do remember of course that I was also the manager of the Sokol outlet after the war?

Naomi: How could I forget? You worked for my oldest brother, David.

Barbara: I have a surprise for you. I have a diary from my youngest brother, Carmen, from the war!

Naomi: I didn't know you could keep a diary [aboard ship] during the war.

Barbara: That's true. When Carmen died in an auto accident, a buddy of his brought it to me. He told me how Carmen would write late at night under a blanket with a flashlight.

Carmen after the war.

Naomi: Tell me about your parents . . .

Barbara: Both of my parents were born in Italy. My father, Vincent Franceschelli, was born in Abruzzi and my mother, Rachel D'Agostino, was born in Salerno. I am the middle child of three. Anthony was the oldest, then me, and last but not least, baby Carmen. We all lived through the great Depression.

Carmen Franceschelli and his niec·

Naomi: What was family life like in the war years?

Barbara: We had a large garden, plus fruit trees. We didn't have a car as most people used buses then. We managed with meat and dairy [ration] coupons. We had huge family dinners during the holidays of 25 or more [people]. On Sunday, Dad insisted we all eat together since during the week we would eat before he got home from his barber shop.

Naomi: Now tell me what Carmen was like as a person.

Barbara: Carmen was very special. He was an outgoing, social young man with many friends. He was very handsome with the typical dark eyes, fine features, and thick wavy hair of Italian men. My girlfriends will agree to all I said here.

Naomi: Tell me more about his likes and dislikes.

Barbara: My Uncle Joe taught him to play the clarinet and saxophone. Carmen loved jazz. He was kind, funny, and talented. What I remember most about Carmen was how good he was to his parents and to his brother and me.

Naomi: When did he go into the Navy?

Barbara: Carmen enlisted as he said, "He wanted to see the world." At the time he was working as a draftsman for the Corbin Screw factory in New Britain. "The Hardware City of the World." This was in 1944.

Carmen Franceschelli is standing in the third row down from the top, the ninth sailor from the left.

U. S. Naval Training Station
Sampson, N. Y.

Company 536
July 30 1943

Naomi: Where did he serve?

Barbara: His basic training was his first time away from home. He was 19 years old. He was assigned to the USS *Hornet*. Technically, he was not supposed to write the diary. In letters to us he talked about how they could count the number of planes that left and then count again the fewer that returned.

Naomi: How did you keep in touch with him?

Barbara: I wrote him almost every day. Mother baked each week and would ship off food to him. He shared it with his many buddies. Mail was their life-line and we knew it.

Notes

[1] Lee W. Merideth, *Grey Ghost: The Story of the Aircraft Carrier Hornet CV-12, CVA-12, CVS-12* (Sunnyvale, CA: Rocklin Press, 2001), 30-31, 37-38.

[2] Ibid., 66-68, 82, 85, and 88.

Credits

Acknowledgments

I would like to thank the Franceschelli family and Barbara Franceschelli in particular, for allowing the publication of Carmen's diary. My cousin Donald Z. Sokol, MD, did the initial transcription of the diary and I am grateful to him for all his hard work. I also would like to thank the enthusiastic and supportive folks at the New Britain Industrial Museum—especially Warren and Lois. And my friends John D. Greenhill and Roanne Baldwin for being my cheerleaders in getting Carmen's story in print and out into the world.

Index

About the Author

Naomi Sokol Zeavin was born in New Britain, Connecticut. She experienced World War II there, growing up as a young girl between the ages of 8 and 12 years old. She remembers rationing and going along with her three brothers to collect items to support the war effort. Her family owned Sokol Brothers Coat Factory, one of many factories that supplied war goods to our fighting troops during World War II. Zeavin graduated Dean Academy in Franklin, Massachusetts, and attended Emerson College in Boston. She has lived in Virginia for over 50 years now. She plays an active role in the local history community, including serving on the Fairfax County History Commission. Zeavin is president of U-R-Unique Video Productions. She welcomes you to visit her Web site http://tiny.cc/naomi .